TESTING TRUTHS

TESTING
TRUTHS
THE DEATH OF FREEDOM

DUDLEY H WILLIAMS

SilverWood

Published in 2013 by the author using
SilverWood Books Empowered Publishing®

SilverWood Books
30 Queen Charlotte Street, Bristol BS1 4HJ
www.silverwoodbooks.co.uk

Index prepared by Neil Manley

ISBN 978-1-78132-079-2 (paperback)
ISBN 978-1-78132-080-8 (ebook)

British Library Cataloguing in Publication Data
A CIP catalogue record for this book is available from the British Library

Set in Bembo by SilverWood Books
Printed on paper sourced responsibly

Contents

Acknowledgements

In his preface, the author Dudley Williams thanks those whose opinion he had sought. His widow, Pat, would like to add her grateful thanks to Lester Taylor and Jeremy Sanders for their encouragement and providing help through the final stages of the publishing process – and last, but not least, to his sons Mark and Simon for their support throughout.

A Few Words About This Book

Dudley Williams was a Professor of Biological Chemistry at the University of Cambridge and a Fellow of the Royal Society. His scientific career began with work on the synthesis of Vitamin D while studying at Leeds University and for which he was awarded a PhD in 1961. After spending two years at Stanford University, first as a Fulbright Scholar and then as a Research Associate. He returned to England to take up an academic position at Cambridge University and was awarded an ScD in 1972.

In the 1960s, Dudley's research interests focused on developments in Mass Spectrometry and he pioneered a significant fundamental understanding of this technique and its application to solving biological problems. In the early 1970s, using Mass Spectrometry, he showed how Vitamin D is transformed into an active hormone that controls the uptake of calcium. He was also fascinated by the application of NMR for the structural elucidation of biological molecules. During his career, he applied both techniques to solving the structure of the antibiotic Vancomycin, work which began in 1969 and took twenty years to complete, and then led to fundamental studies on how molecules recognise each other. Dudley published numerous scientific papers related to his research and co-authored many scientific books.

The seeds of this book were sown approximately thirty years ago in the mid 1980s and arise from Dudley's inexhaustible curiosity about our world. Consequently, from its inception the book has undergone many changes. Initially, he worked on the content whenever he had a spare moment, but it wasn't until after his retirement that he devoted more of his time to finishing the book. While the book is aimed at the lay reader interested in science, those with a scientific training may also find key concepts presented with a fresh clarity. The first part of the book examines many of the fundamental principles that help to understand the universe around us, while the latter chapters are Dudley's personal philosophical views based on these truths.

The book was finished when he was admitted to hospital at the beginning of October 2010 where he continued to work on the completion of the index (but didn't quite complete). He died at the beginning of November 2010.

One of his last requests was that I should publish the book for him.

Pat Williams, 2012

Preface

I have no particular genius. But I am passionately inquisitive.

– Albert Einstein

This book is a brief view of what has come to be over a period of 13.7 billion years; one not simply, collectively and consequentially previously told. In the language of yesteryear, it takes us from bang to man or, in that of today, from forces to females. It briefly summarises the evidence that answers the child's question: "How did people get here?" But, more importantly, on the basis of this and other evidence, it then strives to answer their further question: "Why do people do that?" The answers challenge adults and, looking at the world's problems, are of supreme importance – for the vast majority of the world's populace are either unaware of, or do not accept, many fundamental evidence-based truths. It is to them that *Testing Truths* is addressed.

Largely within the last 150 years, science has provided a great deal of new information about what has made us what we are. The first part of this book is not simply a peek into this knowledge but, much more importantly, a portrait of the principles – a brief and simple description of the rules that have controlled the evolution of the Universe and of organisms, and have determined our behaviour. It is the concise and collective description of these rules and principles that illustrates how time and chance determine what we are. By not shirking fundamental commonplaces, there emerges a synthesis that challenges conventional views of our species – a picture very different from "the free choice between good and evil" descriptions that pervade most of society.

Free will is a falsehood that continues to be peddled, for it is a notion that aids the successful operation of social groups of humans within which behaviour must be controlled. But the evidence indicates that we, like our biological cousins, are biological machines lacking free will; sometimes more robotic than we wish to admit; our actions dictated by a few billion years of environments. Through a deeper understanding of how humanity was forged, perhaps we can gain a better sense of how to forge future humanity – collectively a great and important story.

I thank Colin Beard (Roche Biosciences, Palo Alto, California) for reading the text; and Ray Abrahams, Alec Boksenberg, Chris Tout, Archie Howie, Cahir O'Kane and Oliver Kruger for comments on the parts relevant to their expertise – largely in the early chapters. Many sections have benefited from their comments. But the less conventional views are completely mine; all errors and disputed conclusions are to be placed at my door. Most of all, I thank my wife, Pat, to whom this book is dedicated. Without her tolerance and support, simply put – no book.

Dudley Williams, Cambridge, 2010

0

Truths and Concepts

In Nature's infinite book of secrecy, a little I can read.
– William Shakespeare in 'Anthony and Cleopatra'

Chapter zero. Zero. Nothing. Was there ever "nothing"? We cannot yet be sure of the truth of the matter. And in a book that deals with questions and answers, definitions of truths, fictions, and cause and effect must first be addressed.

Truths and Fictions

Theories should have testable implications which, if proved false, falsify the theory itself.

– Karl Popper

To explore the problems of our origins, and what we are, we need to know which of the things that we learn in life are true or false – and shades of probability between these extremes. The scientific philosopher Karl Popper pointed out that a hypothesis should be testable through experiment. It is then possible to prove if it is false[1] and, if false, must be rejected. If, on the other hand, the hypothesis passes all the experimental tests that have ever been made of it, it is called "a truth".

    Isaac Newton's Laws perform excellently in calculating where a stone will land when thrown into the air with a known velocity, and at a known angle relative to the ground. They apply equally to the motion of the Moon around the Earth. They are wonderfully unifying. However, Albert Einstein showed that, if the object is projected at a velocity approaching that of light, a correction is necessary – under some circumstances, Newton's Laws are bettered. But, for most calculations, they are impressively accurate and, under these circumstances, we regard them as "true". They have told us how to get a rocket to the Moon, and back, with such precision that this exacting task has been achieved. Asked to make a precise prediction, they have passed with flying colours – an awesome vindication. If there is failure in a space mission, it is in the machines or other shortcomings in human planning – not in the scientific truth.

But humans are surprisingly susceptible to the acceptance of falsehoods, especially when they appear in print. So much so that the trait makes for satirical comment:

> *...it is something of a mystery why, with one dubious exception concerning a budgerigar living in Dorking, no bird has managed to become capable of live births.*
>
> – Simon Conway Morris in 'Life's Solution'
> Cambridge University Press, 2003

Who is likely to give the budgerigar story credence? Not those who prefer strongly evidence-based statements because, among billions of observations, no reliable exception has been found. Perhaps those who would believe the headline "Elvis seen Alive and Well on London Bus"?

But the "truths" of science cannot lead humanity to a panacea. First, most problems currently lie beyond such tests – they are simply too complex. The variables may be too great in number and/or change in a complex manner so that the outcome cannot be reliably predicted. But, even here, there are often "probable truths" – one outcome is much more probable than others. In a complex world, the differentiation between probable truths and probable falsehoods is crucial. For where the education system of a society fails its pupils, the adults that they become are likely to make flawed decisions. In turn, the adults pass these flaws on to their children. Educational systems of societies bear a heavy load; for where education fails, societies suffer and conflict often arises.

For very good reasons, when rearing children we attempt to give them an impression of a world that is full of certainties, trust and kindness. But, by their teens, they also need to know that the waters into which they will shortly be released are unfortunately shark-infested; that money should not be borrowed at a *per annum* interest rate of 16% when inflation is only 3%. So obvious, yet people commonly do it, as when they fall into credit card debt; and not trivial, because economies collapse when debt spirals out of control. Any system of education fails its children if it does not teach basic economics to them at 13+. For banks, as for all institutions that seek to survive in a highly competitive world, pedal both good and bad advice, both truths and falsehoods, to their customers. Who can blame the young for being confused when CEOs and senior politicians relate that the financial failures of their systems arose when "we were pulled down

2

by forces outside our control", but say that their successes were due to personal, wise decisions?

The Utopian dream of universal fairness fails because institutions try to survive at the expense of their competitors. In any area of commercial activities, the CEO may be appointed on the basis that he/she will receive a multi-million dollar bonus if profits exceed a benchmark. Gigantic salaries are justified on the basis of attracting only "the best". "The best" may satisfy short-term profits, but they fail to understand long-term and complex issues – hence the collapse of banks. Such are the dangled rewards that probable truths are denied; ethical behaviour a casualty.

It may seem astonishing that many major banks were, only shortly after their collapse in 2008 and a taxpayer bailout, once more justifying large salaries and bonuses to employ only "the best". But the recent "best" squandered the hard-earned savings of others. The lesson is that probable truths are missed not only due to the poor dissemination of wisdom but also buried due to complexity, or denied in the vested interests of sub-groups. Falsehoods are promoted in efforts to separate people from their hard-earned cash. This may all be deemed as "obvious to everybody". But, if it were, then the world economy would not have been brought almost to its knees. Misinformation, deceits and lies are so successful in passing wealth from **A** to **B** that they abound. Later chapters tell why they are part of our condition.

Additionally, "testable truths" cannot be reached in cases that involve strong human emotions. When it comes to sentences for murder, there are no universally accepted "right" or "wrong" answers. And such is the nature of our programmed emotions that even testable truths are often denied. "God created Man in his own image" is a truth to many – despite the clear evidence that *Homo sapiens* was not created but, rather, evolved. Conflict arises where an education based in reason might prevent it. And reasoning requires an understanding of cause and effect.

Cause and Effect

Many women who do not dress modestly lead young men astray and spread adultery in society which increases earthquakes.

– Hojatoleslam Kazem Sedighi senior Iranian cleric
in a sermon 16 April 2010

An "effect" is an observable change to a system, and "causes" bring about the change. Scientists observe effects, and seek and analyse their causes.

In so doing, they should make as few assumptions as possible, eliminating factors that make no difference to an observable prediction (Ockham's Razor). Where science is at its most powerful, the change from a particular cause, or set of causes, is reproducible whenever the change is brought about. So, science is a powerful method because there are determinisms giving rise to reproducibility – successful predictions become possible. But, often in complex systems, it is difficult to know all the causes and/or to know their magnitudes with high precision. Under these circumstances, different outcomes may be observed when the experiment is repeated, and predictions are not reliable.

The scientific method does not lead to the conclusion that adultery is a cause of increased earthquakes. But, even among people who have some education in the scientific method, there is confusion regarding cause and effect. Consider two effects that can be observed and quantified. Where changes in the magnitudes of these two things (*variables*) go hand in hand, the two variables are *correlated*. When a correlation is found between two variables, it is often falsely concluded that one of the variables is the *cause* of the second variable. The false inference of causes from correlations is an amazingly common occurrence, and represents a serious failing in systems of education. Unwary scientists, and journalists of national newspapers, often fall into the trap – and it's easily done.

"Your name may even help you get into Oxbridge" proclaimed an article in *The Times* of London in 2007 – a conclusion which, if proven, would rightly elicit cries of disgust. The data revealed that certain given names cropped up more frequently at Oxford and Cambridge Universities than elsewhere. But "desirable" names vary according to the environments that the parents experience. So, the data show only that students have a varying probability of ending up in Oxbridge according to the differing environments they have experienced. This is in itself a social dilemma but one that is a consequence of the whole social system rather than of the universities alone. The conclusion from this data that certain given names *help* to gain admission to Oxford or Cambridge is as logical as a casual assertion that the name Singh helps to gain admission to the Sikh religion.

But more serious errors may arise. The crime rate dropped markedly in the USA over the decade 1991-2001. From around 1980, policing became more effective – a larger number of officers, and more apprehended people were sent to jail. So, there is a correlation between the falling crime rate in the period 1991-2001 and more effective policing from around 1980. It is easy to assume that the larger number of officers is the major cause of the fall in the crime rate since a crime might be prevented by (i) the sight of an officer (ii) jailing of an offender or (iii) the threat of jail deterring a potential criminal. But is this assumption secure? This is a very important question since the assumption may determine the direction of the social policy of a large nation for many years to come.

Steven Levitt and Stephen Dubner conclude[2] that a major influence was in fact a change in the abortion laws. Abortion was illegal in the US until 1971 when five states – New York, California, Hawaii, Alaska and Washington – made it legal in some circumstances. Then, on 22 January 1973, the Supreme Court extended the right to all American women (in a case known as *Roe vs Wade*). In the year following this judgement, about 750,000 women had abortions (one for every four American births). By 1980, about 1.6 million babies were being aborted every year (one for every 2.25 live births).

Childhood poverty and single-parent household are, in the US, strong indicators that a child will have a criminal future. And the historical record establishes that poverty and "single parenting" are well correlated with an increased probability of abortion. It is therefore argued that, following *Roe vs Wade*, the US population came to lack large numbers of young males who would have had a higher probability of being criminals. Hence, it might be that the facile conclusion – better policing caused the reduction in crime – was only a part of the story, and that abortion was also a cause of crime reduction.

Outraged? Perhaps, but the emotions are often key deniers of truths. Sceptical? You should not be, for it was in precisely those five states that legalised abortion about two years before *Roe vs Wade* that crime began to fall earlier than the other forty-five states (and the District of Columbia). These data are powerfully persuasive that the increased abortion rate was a significant *cause* of the decreased crime rate. It may be currently politically incorrect to infer that single parenting has undesirable consequences for society, but the data point strongly to this conclusion for the described period in the US. Some positive effect of more policing is not excluded.

Sometimes, the *direction* in which causality acts is falsely assumed. An elderly person falls, and is later found to have a broken hip. Thus, in the elderly, falling (one variable) is correlated with a break (second variable). It is easy to assume that the *fall* is always the *cause* of the *break* in the hip. However, the problem of broken hips in the elderly is commonly associated with osteoporosis. So, on occasions, it is a strain experienced in the standing position that causes the hip to break, and the *break* is the *cause* of the *fall*. Cause and effect have been reversed.

Once cause and effect were understood and measured, the scientific method permitted an understanding of the evolution of the Universe, organisms, and behaviour. Despite still large holes in our knowledge, the picture that emerged is an enormous advance on the ignorance that preceded it.

Big Players in the Universe

If I have seen further, it is by standing upon the shoulders of giants.

— Isaac Newton (1642–1727)

Mass (m) is a big player in understanding the Universe. It is the amount of material in an object – the stuff on which gravity acts, and of which we (plus all other earthly organisms) and the visible Universe are made. Given the enormous distances to stars ("distant Suns"), a glance out into the Universe on a clear night indicates that it contains a lot of mass.

Energy (E) is also a big player for, without it, change would not occur, and humans could not exist. Energy is most commonly experienced as the strength of the "punch", or more formally "the ability to do work", carried by a moving mass. Masses such as tennis balls, or automobiles, have energy by virtue of their movement. And for a given degree of movement (velocity v) of the object, the automobile has more energy than the ball because it is has a larger mass, and therefore is more difficult to stop. Energy also arrives from the Sun in the form of photons – "sunlight". Since the visible stars emit vast amounts of energy, a glance at the visible Universe establishes that it contains a lot of energy.

Gottfried Wilhelm von Leibniz, born Leipzig, 1646, got very close to defining the energy of a moving mass: its kinetic energy. Gottfried realised that the kinetic energy was proportional not only to its mass (m) but also, given a velocity v, to $v \times v$ – written as v^2. [Later, the kinetic energy (E) was shown more precisely to be one half of $(m.v^2)$.] If we wish to reduce the chance of being involved in a traffic accident, some basic science is useful. For if the speed of a vehicle is increased by a factor of 3 then, once the brake is applied with a given force, the stopping distance goes up by a factor of 3×3, ie 9. In the same period that Leibniz was active, Isaac Newton related the concepts of mass, force (the strength of a push or pull), velocity and acceleration in his famous Laws. Understanding the motions of the vehicles of the future would not be a problem.

Newton and Leibniz were mathematical giants of their time. They ended up in bitter dispute with regard to their relative contributions in another area of science – the discovery of calculus. The dispute makes for fascinating theatre.[3] Samuel Pepys, as President of the Royal Society accepted Newton's "Principia" in 1687. However, Isaac, as a later President of the Society, was able to pull strings with regard to the dispute over the invention of the calculus. He set up a committee, loaded with individuals who were persuaded to his cause, to make a judgement on priority. Isaac then proceeded to write the report of the committee himself. As a human being, Newton was capable of distortions and, in some ways, he was an objectionable character. But his equations cannot be distorted.

In the transformations that are commonly experienced, mass, which figured so importantly in the minds of Newton and Leibniz, is conserved – it can neither be created nor destroyed. For example, the French scientist Antoine Lavoisier (1743-1794) discovered that mass is conserved in chemical change. He recorded that when water (formerly regarded as an element) is split into the true elements of hydrogen and oxygen, the sum of the masses of these two gases is equal to the mass of water from which they came. An iron cannon ball may partially rust but, in one form or another, all the original mass of the metal ball is still with us – either as unchanged metal, or in combination with oxygen as oxides of iron. Come the French Revolution, Antoine's greatness offered no protection – as a former tax collector, he was considered an enemy and guillotined in 1794.

In the transformations that we commonly experience, energy is also conserved – the First Law of Thermodynamics. If work is done to lift a stone onto the roof of house then the work done is stored in it as potential energy. For if the stone is then allowed to fall from the roof, the kinetic energy it possesses on hitting the ground is equal to the energy put into it upon raising it from the ground to the roof. No net loss or gain – the "books are balanced".

When petrol combines with oxygen to produce the energy that propels a car, the products of combustion contain all the original in-going mass. The energy is released simply by rearranging the bits that make up the mass. Both mass and energy are rigorously conserved. Famously, Lord Kelvin assumed these conservations in an effort to estimate the lifetime of a "burning" Sun made of coal-like material. It was calculated that it would burn out within a few thousand years. Alternatively, if the energy of the Sun were derived from the material within it simply collapsing under its own gravitational attraction, the lifetime could be extended to around thirty million years. Both estimates were peanuts compared with the hundreds of millions of years proposed for the lifetime of the Earth by Charles Darwin and contemporary geologists. Since the lifetime of the Sun and the Earth were very likely to be similar, it seemed that either Darwin or Kelvin was wrong. It was Kelvin who was wrong. But it required Arthur Eddington, using radical new ideas developed by Albert Einstein, to teach us why.

Einstein's story is a classic case that illustrates that fundamental discoveries require extended deep thought. Light (photons) travelling from the Sun to our eyes has a velocity of 300,000 kilometres per second. Einstein imagined himself sitting on a light beam. Would the light, relative to him, then be stationary, or would it still travel away from him at 300,000 kilometres per second? That is a *deep* problem, and a revolutionary solution was required to solve it. Einstein hypothesised that the light would still travel away from him at 300,000 kilometres per second – and you don't get more counter-intuitive than that. His thinking led him to conclude that the velocity of light is a true constant, but that the rate of progression of time is not.

His work, the most stunning of which was published in 1905, led him to the famous conclusion $E = mc^2$, where c is the velocity of light. Since c is a very large number, a little bit of mass (m) can be converted into an enormous amount of energy (E). If sceptical of $E = mc^2$, take on board the existence of atomic and hydrogen bombs. In 2005, the equation was shown by experiment to be accurate to better than one part per million.

Thus, although everyday experience suggests that mass and energy are conserved independently, the universal solution is the conservation of mass/energy. Mass is a very concentrated form of energy. Given particles colliding at extremely high temperatures, mass can be converted into energy, and the Sun appears as an almost eternal fount. For his monumental contributions to science, including much outside the field of relativity, Albert Einstein received the Nobel Prize in Physics in 1921.

The "Big Bang" – A Beginning?

All great truths begin as blasphemies.
– George Bernard Shaw

As time passes, what happens in the Universe is consequent upon what has already happened. So, to understand ourselves, we must start at the earliest point of which there is knowledge, and has consequences in the causal chain leading to ourselves.

That 13.7 billion years ago, the cosmos commenced an expansion from a remarkably dense state is an evidence-based view. We have great difficulties in comprehending such enormous timescales. Take a period of 100 years, which is a long time by human standards. If we write "then another 100 years passed" a sufficient number of times to give a total passage of 13.7 billion years then the repeated phrase would fill approximately fourteen thousand books each of 100 pages in length. And, in this time, humans – and their innumerable fellow creatures – have spontaneously evolved from a chaotic cauldron of particles. This journey of matter and the properties of the emerging systems – as amazing as any science fiction – are the stories told in the following pages.

The matter of the Universe was not only remarkably concentrated 13.7 billion years ago, but also incredibly hot and in the first phase of an expansion that still continues. Conceivably, it was at the time of the "Big Bang" that the matter of the Universe was formed. This possibility immediately poses the question: "From what?" Although there are views, they are currently speculations. Erasmus Darwin (Charles Darwin's grandfather) had got round the problem by proposing a model of an oscillating Universe: a system undergoing repeated expansions (from a bang) and contractions (back into a crunch). Any theoretician entertaining this notion is now obliged to do so within a scenario of the period of each cycle being at least many tens of billions of years. But the current evidence – see later comments on "dark energy" – suggests that a hypothetical "Big Crunch" will not follow the well-established "Big Bang".

So, the "Big Bang" – a name coined by the physicist Fred Hoyle when he wished to ridicule the idea – is our logical starting point. Three powerful pieces of experimental evidence support the concept. First, the weak "radiation background" associated with the "Big Bang" is still with us. It is as though somebody had dropped a hydrogen bomb, and the shock waves were still being felt much later in time – just in the same way as we can detect the ripples which cover the surface of an otherwise placid lake long after the impact of a stone upon the surface.

Detection of the weak background radiation requires a very sensitive receiver. Such an apparatus was built – for other reasons – in the early 1960s at Bell Telephone Laboratories in the USA, and later used by two radio astronomers, Arno Penzias and Robert Wilson. Although they did not immediately realise that the background radiation that they detected was a remnant of the "Big Bang", their observations were a major breakthrough for the theory. They were awarded a Nobel Prize in Physics in 1978.

The second piece of experimental evidence that supports the "Big Bang" model is that it leads to a successful prediction of the relative abundance of the light elements – atoms with small numbers of protons and neutrons in their (nuclear) cores. Prior to the observations of Penzias and Wilson, George Gamow had put forward a theory for the formation of such elements – deuterium, helium and lithium. He proposed that they could have been synthesised from the hydrogen nucleus when the Universe was in a relatively dense state – only a few minutes after the "Big Bang". It was the high density and temperature at that time which permitted the collisions between the particles (protons and neutrons) that make up nuclear cores. But the formation of even bigger cores, to make possible the birth of the heavy elements (big nuclear cores), could only occur much later – when protons and neutrons were in collision for very long times in the very dense interiors of stars.

George Gamow was a man with great senses of imagination and humour. Through the essentially artificial inclusion of the name of his friend Hans Bethe in one of his key papers, it appeared as authored by Alpher, Bethe and Gamow, and became known as "the alphabetical article". When, at the end of his life, he spent a period in my own college (Churchill) in Cambridge, I saw him retiring to bed with a bottle of whiskey. He had decided that his life's work – which included a highly popular book on the consequences of relativity[4] – was done.

The third piece of evidence to support the "Big Bang" theory is that distant galaxies are receding from us at enormous speeds. Imagine that you are a participating particle in this expansion. If you look at another particle, you will see that it is receding from you. The further it is from you, the faster it recedes – exactly what is seen when the visible Universe is examined from Earth.

It was the American astronomer, Edwin Hubble, who realised that galaxies are rushing away from each other at a rate proportional to their inter-galactic distance. We know this because the more distant a galaxy is from Earth (which is in the Milky Way galaxy), the more the wavelength of the light which is reaching us from it is longer than expected – the famous "red shift". This is due to the Doppler effect, which is familiar in everyday life – the sound of the horn of a receding car is lower in pitch than when the car is approaching. When the car is receding, fewer peaks and troughs of the sound wave arrive per second at the ear than when the car is approaching. Thus, in the case of the receding car, the frequency of the sound is lower. An analogous effect operates when we observe the distant galaxies: because they are receding, fewer peaks and troughs of the light wave arrive per second at our detector than would otherwise be the case. The apparent frequency is decreased.

Thus, the theory of an expanding Universe commencing from a very dense and hot state is not fanciful. It is based on lots of solid evidence, and has survived a number of challenges. That the expansion started many billions of years ago is known not only from the red shifts observed for galaxies at various distances, but also from features of the microwave background radiation. The good agreement between the two methods gives further strong support to the theory. It eclipsed a rival theory, put forward by the physicists Fred Hoyle, Tommy Gold and Hermann Bondi, which proposed that the Universe is in a "steady state".

Those outside the field of cosmology tend to imagine the expanding Universe gradually filling more *pre-existing* space as it spreads out in all directions. However, Einstein's reasoning taught us that the Universe does not expand to fill pre-existing space. Rather, space, time and matter are all intimately interlinked, and one does not exist in the absence of the other two. Since space can only be associated with time and matter, the Universe has no edges – the "Big Bang" occurred in all space. An excellent two-dimensional model of the expanding Universe is the surface of a balloon as it is blown up. Tiny spots on the balloon surface represents matter, and the unmarked area represents space – expanding as the balloon expands. In this model of the Universe, space (represented by the surface of the balloon) is finite, but without boundaries. In removing this dilemma, a beautiful analogy to the problem facing thinkers in the Middle Ages is found: what happens when we get to the edge of the world – do we drop off the end? The answer to that problem is that, although the world is finite, it has no end to drop off.

We have no way of testing whether the Universe is finite or infinite. The human brain finds an infinite Universe impossible to intuit – a system that goes on forever, and one in which the existence somewhere of a carbon copy of yourself is not of almost zero probability but instead a certainty. But even the observable Universe is of a size that challenges the limits of our imagination. Distant stars seem to be in the same place even when observed from the opposite ends of the Earth's orbit round the Sun (ie when using observations made six months apart). When such stars are used as a point of background reference, stars that are much closer (in cosmic terms) are observed to change slightly the angle that they subtend to the observer positioned at the opposite ends of the Earth's orbit. By measuring this change in angle, trigonometry can be then be used to measure their distance. For observations made from the surface of the Earth, this method (using "parallax") is useful up to about 2 thousand trillion miles (about 300 light years) – and up to about 20 thousand trillion miles if the observations are made from telescopes in space. But relatively few stars are such near neighbours.

Other methods are required to "step up" to the measurement of larger distances. Given a candle then its distance can be estimated from how its brightness is attenuated with increasing distance. The Universe contains objects that are of roughly known brightness ("standard candles").

Therefore, by analogy to the case of an earthly candle, observation of these "standard candles" permits a calculation of their distances. By using supernovae (very bright events resulting from the explosion of stars), observations from the surface of the Earth allow distances of up to about 2 billion trillion miles to be determined – and even larger distances if telescopes in space are used.

In a finite Universe, all matter must have been in close proximity to all other matter at the time of the "Big Bang". The red shifts of the most distant galaxies indicate that they are receding from us at close to the speed of light. So their distance from us can, given some simplifying assumptions, be estimated. The maximum distance corresponds to the matter receding at velocities comparable to that of light (186,000 miles per second, 300 million metres per second, or about seven times around the Earth in a second) for a period – in round terms – of 10 billion years. This allows a rough calculation of the maximum distance of the most distant observable galaxies as $186,000 \times 3,600 \times 24 \times 365 \times 10 \times 1,000,000,000$ miles; that is, at a distance of about 60,000,000,000,000,000,000,000 miles. To get there would require a journey of one trillion miles repeated 60 billion times; the Universe demonstrated as a piece of real estate, the size of which is difficult to comprehend.

It may initially seem curious that large stars have shorter lifetimes than do relatively small ones, such as the Sun. In a star of very large mass, the pressure due to gravitational attraction is enormous. Atomic nuclei are then relatively rapidly "pressed" into more stable forms – the star burns its hydrogen fuel very rapidly. A star ten times as big as the Sun burns out in a mere 20 million years, whereas the Sun itself will have a lifetime of almost 10 billion years.[5]

It follows that if there exist clusters of stars in which all the members that remain visible have masses less than that of the Sun then the cluster is likely to be more than 10 billion years old. In fact, there exist globular clusters of stars where the members have only less than 0.7 times the mass of the Sun. Apparently, the more massive members of the cluster suffered their demise by "burning out". So, these clusters are likely to have lifetimes of at least 11 billion years – further evidence that the Universe is a very old place.

Because of the billions of years that light from very distant galaxies has taken to reach us, they are seen where they were many billions of years ago. They are observed now in the state in which they were billions of years before the Earth had even formed – and even longer before the evolution of the Earthly life that now observes them. The Universe is a wonderful place in terms of reporting to us, at this very minute, details of its early history; or, as the cosmologist Hermann Bondi was fond of saying, "The Universe is a wonderful laboratory."

Does current experimental evidence suggest that the Universe will continue its current expansion forever, just grind to a very gradual halt, or gradually halt and then contract? The currently available data support the idea that the Universe will continue its current expansion forever. But, what appears most clearly is that there is a relatively fine balance between the three possible outcomes. The factors involved can be understood from the analogy of a stone thrown up from the surface of the Earth.

When the stone starts its upward path, it has its energy of upward motion – kinetic energy – given to it by the thrower. As it proceeds upwards, it slows down because of the pull of the Earth. Work is done against gravity. But where is the kinetic energy going? It is passing into potential energy such that, when the stone has reached its maximum height, all the kinetic energy has passed into potential energy. We know the potential energy is there, because when the stone falls again and hits the surface of the Earth, the potential energy has all been converted back into kinetic energy. As the stone passes through its "up-and-down" journey, the sum of the kinetic and potential energy is constant.

Think of all the possible throws that might not be enormously strong, and result in the stone falling back to Earth. Think of all the possible throws that might be incredibly energetic, and result in the stone readily zooming into outer space. Finally, think of the low probability of making a throw where the stone would be very close to stationary just at the point when the stone was so distant from the Earth that the Earth's gravitational attraction for it was close to zero. This last finely balanced case is that found for the Universe. It is as though a stone had been thrown from the bottom of a deep well with the right amount of energy to bring it to a halt at the rim of the well.[6]

This "balancing act" has very non-trivial consequences. For if the expansive forces of the Universe had been much greater than the restraining forces then everything would have been dispersed without aggregation. Alternatively, had the restraining forces been much greater, there would long ago have been a collapse back to a dense state. In either of these scenarios, the Earth and its constituent organisms would not have existed.

Dark Matter

Until relatively recently, it had always been assumed that the mass of the Universe was simply associated with is visible objects. But when Newton's

Laws were used to test this assumption, it turned out that there is more celestial mass than can be seen. But how can the mass of celestial objects be estimated? It can be done through the application of Newton's Laws.

Consider determination of the mass of the Moon, circulating round its Earthly core. First, we need to know the mass of the Earth – which constrains the Moon in its orbit. This is calculated from the force with which the Earth attracts an object of known mass at its surface. The fact that the attraction between the masses of Moon and Earth (due to gravity) must be just big enough to hold the Moon in its orbit is used next. This constraint allows a calculation of the mass of the Moon.

Newton's Laws can similarly be applied to understand the motions of galaxies – vast collections of stars – that rotate. Among the around 100 billion galaxies that make up the visible Universe (which, on average, may contain around 100 billion stars), there are spiral galaxies. In these spiral galaxies, clouds of hydrogen gas rotate round the centre with a velocity that can be estimated. When the arithmetic for these rotating spiral galaxies is done using the same principles as for the Earth/Moon system, the estimated mass of the galaxy is greater than the mass of its visible part (stars) by a factor of about six.

Other evidence also points to the conclusion that the vast majority of the mass of the Universe is not giving out light, or related electromagnetic radiation. It is concluded that the missing mass is not constituted from the normal constituents of atoms. As a consequence, we humans are not made out of the same stuff as most of the Universe. "Dark matter" is a problem – and a big one.

But the problem of "dark matter" should not deflect us. The expansion of the Universe (Ch 6-9) during 13.7 billion years was one of the great discoveries of the twentieth century. It is supported by much hard data. During the expansion, due to the work done against gravity, the particles with mass slowed down. Since the more slowly moving particles are effectively cooler than they were at times close to the "Big Bang", they have tended to aggregate into larger clumps as time has passed – thus accounting for the formation of stars and galaxies. As time passed, a more interesting Universe was generated – as next explored.

1

A Cooling Universe

Out of the Fire

A wise man proportions his belief to the evidence.
– David Hume (1711-1776)

Air is cooled when escaping from the nozzle of a tyre. In a similar way, the Universe cooled upon expanding. As it did, the very small entities from which it was initially constituted joined together to form larger aggregates – atoms and then molecules. With the passage of more time, the very large aggregates of stars and galaxies were formed; and, at even later times, large molecules and then the enormous aggregates of molecules that we call organisms. All these constructs were successively "born"; each took on its own characteristics, and is given its own name. The aggregates that were formed relatively early in the evolutionary game are used to describe the structure of the Universe; those formed later upon Earth to describe species.

The spontaneous aggregation is due to attractions – forces – between the smaller particles. A good analogy is found in the behaviour of a set of ball bearings in a concave dish. In the absence of agitation, the ball bearings behave as though their temperature is 0°K (-273°C). They have no kinetic energy and cluster together at the bottom of the dish. But, given sufficient agitation (energy), the cluster is disrupted. The cluster represents an aggregate – usually described as particle with a new name – and the individual balls represent the constituent parts. Thus, when the smaller particles are given a lot of energy, they naturally exist as discrete entities. But when the kinetic energy present in the cluster is less than that necessary to overcome the attraction between the smaller particles then the large particle – the cluster – is the entity that exists.

A high temperature means that particles are banging together with large energies. Under these conditions, smaller particles are more likely to exist rather than the larger aggregate. In the light of these principles, consider an expanding Universe – a very large assembly of such particles which are initially able to interact with each other through collisions.

During the expansion, the particles gradually lose the energy with which they bang together because of the work that must be done against the gravitational attraction experienced by matter. The kinetic energy of the Universe falls as it expands. Consequently, its initial constituents cool and gradually join together – due to a number of attractive forces between them – to give aggregates.

In the expanding and cooling Universe, there are characteristic temperatures at which the more complex aggregates appear. The temperatures and timescales for production of the various aggregates are given in Table 1-1, and are approximate only (largely based on those presented by Steven Weinberg in his famous book 'The First Three Minutes'[1]). More precise values are relatively unimportant – what matters is the overall picture and the principles behind it.

Table 1-1
Some Particles Found in the Universe

Particle or System	Constituents	Temperature (°C) at which the Particle (System) is unstable[a]	Time for first formation of Particle, or System[b]
Electron	(Fundamental particle)		
Proton	Quarks	10^{13}	1 millionth of a second
Neutron	Quarks	10^{13}	1 millionth of a second
Atomic Nucleus	Protons and neutrons	10^9	1 minute
Atoms	Atomic, nuclei, and electrons	10^4	300+ thousand years
Molecules	Atoms	$10^2 - 10^4$	100–1,000 million years
Stars and Galaxies	Atoms, nuclei, and electrons	$10^4 - 10^7$	100–300 million years
Heavier Elements	Atoms		A few billion years
Planet Earth	Atoms and molecules		ca. 9 billion years
Living Organisms	Organic Molecules	10^2	ca. 10 billion years (on Earth)

[a] 10^9 or 10^2 mean, respectively, 1 with 9 zeros after it, or 1 with 2 zeros after it. Thus, $10^4 - 10^5$ means 10,000 to 100,000, and the implication is that atoms break up into their constituent nuclei and free electrons at temperatures greater than this range.
b time since the "Big Bang".

What are the attractive forces – the Universal glues – that give rise to the aggregates of Table 1-1?

The Universal Glues

Physicists have discovered four forces, the glues of the Universe, which bind the particles together. The *strong nuclear force* holds quarks together to form protons and neutrons, and also holds protons and neutrons together to form atomic nuclei (Table 1-1). The *weak nuclear force* is, as its name implies, much weaker than the strong nuclear force. These two forces are described as "nuclear" because they act over such short distances (at which both are remarkably strong) that they are of consequence only within the very small nuclei of atoms.

As the Universe expanded and cooled, the two nuclear forces came into action before the other forces because they are by far the strongest at very small distances. The next force to come into operation was the *electromagnetic force*. It pulls together particles of opposite electrical charges (+ and -), and thus accounts for the fact that, given a low enough temperature, electrons (which have a negative charge) cluster round nuclei (which carry a positive charge). Among the four forces, the electromagnetic force is unique in having a repulsive equivalent, which opposes aggregation. The repulsion occurs where both the interacting charges are negative (- and -) or both positive (+ and +).

The description of this force as "electromagnetic", rather than simply as "electric", stems from the pioneering work of James Clerk Maxwell. His theory predicted that when a magnet moves, or an electric current changes, a wave of energy spreads out into space. He calculated the speed of the waves – electromagnetic waves – and it turned out to be the same as measured for the speed of light. The full triumph of his theory was realised only over twenty years after it was proposed, when Heinrich Hertz generated electromagnetic waves by means of a spark discharge. James's ideas were epoch-making for it is the existence of electromagnetic waves that permits radio and television. Think of him when using a mobile to call, *via* satellite, halfway round the globe.

Many physicists regard James Maxwell's contributions to their subject as being comparable to those of Albert Einstein and Isaac Newton. James, born in Edinburgh in 1831, died at only forty-eight. As a boy, he showed remarkable curiosity. When observing phenomena for the first time, he was inclined to ask "And what's the go o' that?"[2] Like Einstein and Newton, he wondered how the Universe worked.

The fourth force is gravity – any particle with mass attracts any other particle with mass. Evidence is now accumulating for a fifth force. Although most commonly referred to as *dark energy*, we may also call it the *dark force*. The term "dark" correctly implies that we know little about it, although it is possible that it is a variation on the gravitational force – but repulsive rather than attractive.

What is the evidence for dark energy? The expansion of the Universe was, until very recently, believed to be slowing down. This was the expectation on the basis that, with the passage of time, the work done against the gravitational self-attraction would gradually reduce the kinetic energy of the expansion. Recent measurements have, in fact, provided evidence that the rate of expansion of the Universe is *increasing*. These measurements suggest that there may be a fifth force that, rather than holding bodies together in the manner of gravity, acts to push them apart – an anti-gravity, or "gravity turned on its head". In its effects upon the rate of expansion of the Universe, it began to win against gravity about 5 billion years ago. The effects of this dark force (dark energy), like those of dark matter, seem clear; their origins completely obscure.

Nuclear forces account for the existence of nuclei; the electromagnetic force for the existence of atoms, molecules and their aggregates; gravity for galaxies, stars and planets. But we have no idea why there are four forces – rather than any other number – holding together the material of the Universe. Nor have we any idea why the four forces have their relative magnitudes.

But these magnitudes lead to some interesting consequences.[3] Electrical attraction between the electron and the proton in a hydrogen atom is 10^{39} times larger than their gravitational attraction. So we can neglect gravity in the description of the atom. But gravity becomes dominant over electrical forces at the equivalent of 10^{57} proton masses. At these aggregate masses, and greater masses, the structure of atoms is destroyed. And then, aided by the large temperature rises that are consequent upon gravitational compression, gravity is strong enough to reorganise the structures of atomic nuclei – hitherto fixed by the strong nuclear forces. Changes in the structures of atoms result in more stable nuclear arrangements being generated, and lead to the liberation of light and heat. Stars must be very big (at least 10^{57} proton masses for gravity to start to play havoc with the other forces), and necessarily become very hot when burning their nuclear fuels.

On Earth, the electromagnetic force is important over distances from about 10^{-10} metres up a few metres. Standing at arm's length from a neighbour, if each of you had one per cent more electrons than protons, your mutual repulsion would be enough to lift a mass equal to that of the Earth.[4] So, unsurprisingly, on a dry day, a brush transferring even a few electrons from hair causes hair to rise to the brush's bait. But, in our daily world, chunks of matter normally contain essentially equal numbers of positive and negatively charged particles, and the electromagnetic force is humbled. So, when conversing with a neighbour, there is normally no experience of an electrical force between the participants. Our existence depends on all four of the glues, but we are normally only aware of gravity in the routines of the day.

Physicists strive to find a "theory of everything" to explain, among many things, why there are only these few forces, and why they have their relative magnitudes and distance dependencies. The theory is currently conspicuous through its absence. Unification has in recent years been sought through "String theory".[5] It is mathematically appealing, but has not yet led to predictions that can be tested through experiment. Rather, it is an ingenious speculation that has not yet gained wide support in the community of physicists. Even if the "theory of everything" ever appears, it would not live up to its grand title. Physics would be enormously unified, but the complexities of life would lie beyond its tentacles for it builds astonishing objects from a simple Lego kit.

The Universe's Lego Kit

A collection of facts is not science any more than a pile of bricks is a house.

– Jules Henri Poincaré

The ancient Greeks defined the atom as the smallest, indivisible unit of the Universe. When the particles that are called atoms today were first discovered, they seemed to fit the bill. But, inevitably, the question "But what's inside an atom?" was posed and they were subjected to an energetic battering. Their disgrace was to succumb and split – they failed to satisfy the Greek definition.

So, excepting the mysterious dark matter, what are the current views on the Universe's Lego kit – particles that are regarded as "fundamental" or "elementary" (not known to have any substructure)? The "Standard Model" incorporates sixteen particles, and tells us protons and neutrons are made up of quarks. The sixteen include not only the elementary particles from which matter is constituted, but also a set of particles that transmit three of the four forces (gravity has so far refused to join the club). The Standard Model currently stands at a limit where it meets the wall of presently obscure "ultimate causes". It is sometimes said that, to find answers to "the 'why' of ultimate causes", one asks a priest. But religions were wrong in their descriptions of "creation".

Imaginative guesswork is no substitute experimental test and, in any event, "ultimate causes" become unnecessary if the universal system is of infinite lifetime.

Who were the "prophets" whose thoughts and experiments taught us much about the Lego kit of the Universe, and the forces between them? The idea of the transmission of forces through particle exchange won the 1965 Nobel Prize in Physics for Richard Feynman (shared with Sinitiro Tomonaga and Julian Schwinger). His model was an astonishing insight by a great scientist – one who questioned prior assumptions, and often found them wanting. Murray Gell-Mann received the Nobel Prize in Physics in 1969 for his fundamental contributions to the theory of elementary particles – and used the term "quark", coined by James Joyce in his novel *Finnegan's Wake*. The Italian physicist Enrico Fermi (Nobel Prize in Physics in 1938) put forward the first theory of the weak nuclear interaction.

Sheldon Glashow, Abdus Salem and Steven Weinberg then made further fundamental advances for which they jointly received the 1979 Nobel Prize in Physics. Their considerations led to the idea that, at sufficiently high energies, the weak and electromagnetic forces would be unified. By smashing particles together at enormous velocities in the European laboratory for particle physics at the CERN facility in Geneva, the conditions believed to exist one ten thousandth of a microsecond after the "Big Bang" have been simulated. The effective temperature in these experiments is more than four trillion times as great as room temperature. At these incredibly high energies, the theoreticians dream became reality: the weak nuclear force and the electromagnetic force did indeed merge into a single "electroweak force" – one up for theory. Simplicity and unification appear as we go backwards in time.

Astonishingly complex as humanity is, it is constructed from a relatively simple kit. "Up" and "down" quarks give us protons and neutrons, which then bind together to give atomic nuclei. These nuclei then combine with electrons to give atoms. Photons – the particles without mass that transmit radiation – pass energy between both proximate and distant objects, enabling humans to watch the Olympics in China while drinking beer in Johannesburg. This scene is an amazing journey from the "Big Bang".

So, given the four known forces, and using a description involving only protons, neutrons, electrons and photons, a simplified story of the journey of matter can be told. It is a story that eventually leads to stars, planets, rocks, sea, air, iron, mercury, bacteria, mushrooms, crocodiles, eagles, baboons, milk, beer, pizzas, diamonds, supermarkets, cars – you name it. Oh, and humans – for a human is a self-organizing system of about 10,000,000,000,000,000,000,000,000,000 protons – and a similar number of electrons and neutrons. Some Lego kit! And in building it, let's start with the birth of the atomic nucleus.

The Birth of Atomic Nuclei

It is estimated that about one hundredth of a second after the start of the "Big Bang", the temperature was about a hundred thousand million degrees Centigrade (10^{11} °C). After about one tenth of a second, it had dropped to about thirty thousand million degrees Centigrade (3×10^{10} °C); and, after three minutes, down by a further factor of thirty – to around one thousand million degrees Centigrade. At this temperature, the energy with which the particles hit each other was considerably less than three minutes earlier.

At these early times in the Universe's birth pangs, there were plenty of colliding protons, neutrons and electrons. Due to the mutual attraction arising from opposite electrical charges, would it be a proton and an electron that would first aggregate? If so, a hydrogen atom, with the electron "orbiting" round the much more massive proton – originally envisioned as analogous to the Moon orbiting round the Earth – would be a very early occupant of the Universe's space. Or, due to the strong nuclear force, would neutrons and protons first combine to produce a naked atomic core?

It is because the strong nuclear force is amazingly powerful at very small distances, and stronger than the electromagnetic force, that small atomic nuclei were born before an electron stuck to a proton. Small atomic nuclei formed early in the history of the Universe – they were born about a minute after the "Big Bang". The formation of big atomic nuclei, containing lots of protons and neutrons, would have to await the formation of stars; bundles of joy yet to come, for organisms are not possible without them.

The Birth of Atoms

As the Universe cooled further, the weaker electromagnetic force came into play. It resulted in the formation of atoms – miniscule spheres – in which electrons are bound to atomic nuclei in numbers that cause the overall electric charge to be zero. The number of bound electrons is equal to the number of protons in the nucleus.

The attractive electromagnetic force – seen in everyday life when a recently used comb attracts paper or hair – had to wait for a few hundred thousand years to exert its influence in the formation of electrically neutral atoms. The exploding Universe had had to cool down a great deal. Atoms are around one hundred millionth of a centimetre in diameter. It is easy to regurgitate diameters of these atoms and the comparable diameters of the small molecules that were later derived from them. But what can convey an appreciation of their dimensions and of their implicit vast numbers? The nitrogen molecules of air are very stable and have a long lifetime. Assume that the nitrogen molecules in the last breath of Julius Caesar, exhaled some 2,000 years ago, are now distributed uniformly throughout the atmosphere. Then, with each inhalation, we take in nitrogen molecules that Julius exhaled in his last. Inhale Caesar.

The evidence for the existence of electrons, protons, neutrons and nuclei came from late nineteenth, and twentieth, century science. It was Joseph John (JJ) Thomson who discovered the electron in his Cambridge laboratory in 1897. For this and other work, he was awarded the Nobel Prize in Physics in 1906. It all seems a long time ago but, on arrival in Cambridge in 1964, I was given the job of examining for the University entrance paper. The then chairman of the examiners told me that his predecessor had been "JJ" who, late in life, could be heard to click together his artificial teeth. It is told that his research group, when posing for the annual group photograph, would ask him to sit down before he was allowed to hold the famous apparatus used in his great discovery.

Ernest Rutherford and his colleagues discovered the nucleus. They fired high-energy α-particles (produced by radioactive decay) into a piece of gold foil. They observed that, although the vast majority of the particles went straight through the array of gold atoms, a few of the particles bounced straight back. The results initially astonished them, for the vast majority of the gold atoms appeared – to their projectiles – to be empty space. But, for a few of their projectiles, it was as though shells had been fired at a battleship and a few of the shells had bounced back. Rutherford, born in New Zealand in 1871, carried out his main research in Montreal, Manchester and Cambridge. From his data, he concluded that the atom contains, at its core, a tiny central nucleus. In his model, the much lighter electrons circulated round the core – in a manner analogous to the manner in which the planets orbit round the Sun.

We now know how tiny the atomic nucleus is. In a crude two-dimensional model, the atom (in truth, a sphere) can be imagined as the surface of an athletics arena, with circular running tracks of increasing diameter in which the electrons are constrained. The atomic nucleus is then represented by a ball bearing of around one centimetre diameter at the centre of the athletics arena. Although the electrons constrained in these tracks are now known not to be circulating, and the tracks themselves to be diffuse, Rutherford's model laid the foundations for the modern view of the structure of the atom. He received the Nobel Prize in Chemistry in 1908, in part for his investigations into the disintegration of the elements. And, although he was not to know at the time of his epoch-making discovery, there was an irony in it. The very α-particles that he had used as "bullets" were themselves a nuclear core – a helium atom stripped of its two surrounding electrons. It required a nucleus to discover the nucleus.

It might be considered most logical to follow the birth of atoms by details of the birth of molecules from the combination of atoms (eg H + H $\rightarrow$ H$_2$, where H$_2$ is the chemist's representation of a molecule of hydrogen gas). However, in the journey to life, it is the evolution of organic molecules that is key. And they could only follow the birth of stars.

The Birth of Stars

The particles that would eventually forge organisms had emerged from the fire of the Big Bang. But they next fell into the frying pan of stars. Why?

The force of gravity complicates the idea of a continually cooling Universe. When very large clouds of hydrogen gas condense due to gravitational attraction, a vast amount of heat is generated – as heat is generated when a massive object is repeatedly dropped on to a surface. Additionally, since the force of gravity is in a sense a catastrophic force – the bigger the clump, the bigger the force – the atomic nuclei inside the larger aggregates were under such enormous forces, and at such high temperatures, that they began to rearrange into more stable arrangements. The energy released in these rearrangements, and from gravitational collapse, causes the temperatures of stars to rise to many millions of degrees Centigrade.

Thus, although the overall expansion of the Universe causes cooling, this is followed by local condensation with the formation of extremely hot stars – crucial for the evolution of organisms. First, because it is within stars that the heavier elements, found to variable extents in organisms, are formed. Second, because gravitational collapse, and the reorganisation of nuclei into more stable arrangements, keeps stars sufficiently hot for light to pour from them. And life on Earth is dependent upon the arrival of light energy – photons – from our own star, the Sun.

A slowing of the rapidly moving atoms in the early Universe was necessary before they could aggregate into clumps. But why was there a *long* wait – around 100 million years – before they did so? The answer lies in the fact that you do not find your neighbour attractive. A human is an incredibly large number of atoms working in concert. Take a million (10^6) atoms and then take a million more for each atom that you have already. You now possess 1,000 billion (10^{12}) atoms, enough to make one millionth of a billionth of a human being. In other words, a human consists of around 10^{27} atoms. Yet one human does not significantly attract another *via* the force of gravity.

But the Earth is sufficiently massive (6×10^{24} kilograms) to attract humans, and is in fact a major determinant of our size. And so it is with the Universe. Once the initial crowd had been dispersed by the awesome kinetic energy of the Big Bang, it took a long time to assemble new ones. But, as with a crowd of rubbernecks, once the new crowds of atoms got going, they found it increasingly easy to suck in others. So, 100 million years after the Big Bang, stars were born from the condensation of enormous clouds of hydrogen and helium gas. The progression of aggregation in the Universe (atomic nuclei –> atoms –> stars) is a consequence of the fact that the forces involved lie, in terms of their strengths at small distances, in the order: strong nuclear > electromagnetic > gravitational.

Forging Building Bricks in Stars

The building bricks that would eventually permit the coming of organisms are the elements. An element is a substance composed of only one kind of atom – each characterised by the number of protons in its nucleus (equal to the number of electrons surrounding the nucleus). So, the larger the number of protons that can be contained within a nucleus, the greater the variety of the possible building bricks. Around 100 elements, containing from one to one hundred protons in the nuclei of their constituent atoms, are known on Earth.

The reason that we have only about 100 elements is that nuclei larger than this are unstable – too many protons near together, and therefore too much electrical repulsion between them to form a stable entity. Thus, the number of elements is determined by the relative magnitudes and distance dependences of the strong nuclear force (binding the nucleus) *vs.* the electromagnetic force (destabilising the nucleus). The kit for making organisms is thereby limited in size. If the electromagnetic repulsions between the protons within nuclei were much stronger relative to the strong nuclear force, there would be fewer elements, and calcium would not have been around to be utilised later in making the bones of the vertebrates. But calcium was around, and evolution is opportunistic.

Why is it that the heavier elements (the ones with bigger nuclei) could be forged in stars? When the first stars formed, the elements they contained had small atomic nuclei – hydrogen (lots) and helium (some). This is because protons and neutrons encountered each other only occasionally in the early Universe, and there was only limited fusion to form helium. The formation of much larger nuclei requires collisions of large numbers of protons and neutrons. This, in turn, required very high concentrations of helium and very high temperatures. Stars are the pressure cookers in which the elements with big nuclei are made.

As seen in the preceding section, stars can initially heat up due to gravitational collapse. When the temperature inside a star reaches greater than 10,000°C, tiny nuclear cores begin to fuse together to give larger nuclei. As on a hot night in a densely populated city, there is much jostling together, and new combinations are born. The combination of two protons (two hydrogen nuclei) and two neutrons gives rise to a helium nucleus. The helium nucleus is more stable than the four separate particles from which it was formed and, in this process of nuclear fusion, some mass (m) is converted into energy (E) according to Einstein's equation $E = mc^2$. This energy (the main source of energy release in the Sun) causes stars to heat up more, and then they are – rather like electric light bulbs – "switched on". After their veil of surrounding placental dust has been swept away by the wind of energetic particles emanating from within, their light starts its journey out into the Universe.

If the fusion to form helium produces a more stable product then why is it difficult to bring about? An analogy can be found in the behaviour of water in a mountainous region. A lake may exist at 20,000 feet, and energy may be derived from it by allowing its contents to flow down to 16,000 feet. But if the only way out of the lake is a pass at 21,000 feet then a lot of energy (pumping upwards) must first be put into the system. Gravitational collapse initially provides the energy to get over the hydrogen to helium pass. There is then a large release of energy due to this fusion.

If the fusion process could be achieved by human technology, the problems of our energy needs would disappear at a stroke. The problem is one getting over the hill – of generating an extremely high temperature in a spatially constrained and high concentration of hydrogen. In the 1980s, a team of USA and UK researchers claimed that fusion had been achieved without the use of an extremely high temperature ("cold fusion"). The lack of requirement of a high temperature should have given them pause for thought, especially since the claim, if true, would have been of enormous importance – cheap energy. Subsequent "test and check" approaches established, as many already suspected, that fusion had not been achieved.

A model for the synthesis of the heavier elements – completion of the kit necessary for the production of organisms – could be proposed following the observation of the explosive death of a star in 1987 (supernova 1987A).[6] This star appears to have been formed a mere 11 million years ago, and had a mass about 18 times bigger than the Sun. As a consequence, it had a relatively short life (chapter 0). After about 10 million years, sufficient of the hydrogen fuel had been used up to allow a cooling of the core region – the frying pan was cooling down.

But, as the internal particles moved less violently, the inner part of the star contracted and became much more dense. This caused a counter-effect – the contraction of a large mass in a small volume unleashed enormous gravitational forces. As a consequence, the temperature rose to about 200 million degrees – sufficient to ignite helium (with its two protons and two neutrons in its nuclear core) as a nuclear fuel, and to pressurise it to the more stable nuclei of carbon (with six protons and six neutrons in its nuclear core) and oxygen (with eight protons and eight neutrons in its nuclear core) within only 1 million years. Such cycles were repeated, carbon being next to "burn" at 740 million degrees to give still larger nuclei. New bits and pieces were being cooked up.

Extend the analogy of water falling down a mountain – further energy is not available once the water has dropped to the lowest available level. In the case of nuclear fusion, this corresponds to getting to the most stable nuclear arrangement. This does not simply correspond to the biggest nucleus – as it would if only the strong nuclear binding force were involved. There is a counter-influence. For, as the nuclei get bigger, they contain more protons. The positive charges on these protons are mutually repelling; and the more protons there are, the more powerful is this repulsive force. Once the nuclear core is bigger than that of iron (nuclear core of 26 protons), the increasing electrical repulsion between the protons dictates that the optimum nuclear binding energy has been passed. The stability of the iron nucleus accounts for the relatively high abundance of this metal in the Earth – it forms the principal constituent of the central core.

So, once the Universe had cooled sufficiently for the formation of atoms and then stars, the evolution of the system towards more complex units continued within stars. More complex atoms were born in stars for similar reasons that new flavours are forged in frying pans – heat is a catalyst for making new combinations. Much of our understanding of the nuclear conversions that take place in stars is due to the physicist Hans Bethe. Born in Strasbourg in 1906, he lost his position at the University of Tubingen at the advent of the Nazi regime in Germany. He migrated to England in 1933 but, in 1935, was appointed to Assistant Professor at

Cornell University, in the USA, where he remained for the rest of his academic life. His discoveries won him the Nobel Prize in Physics in 1967.

The names given to the elements successively containing one to ten protons are *hydrogen*, helium, lithium, beryllium, boron, *carbon*, *nitrogen*, *oxygen*, fluorine, and neon. The four in italics are the four main constituents of life. Of these four, all but hydrogen are made in stars. But how do they and their even bigger nuclear brothers – necessary for the production of life as we know it – get from star to planet Earth?

The Death of Stars

It is estimated that when the "supernova 1987A" star (as already seen, a star with about 18 times the mass of our own Sun) started to collapse under gravitational attraction, the collapse continued until the core density exceeded 300 trillion grams per cubic centimetre.[6] This density corresponds to 1,000 loaded oil tankers, each of a mass of 250,000 tons, being concentrated into a volume about the size of a sugar cube. At this point, according to one speculative model, the inner part of the core rebounded and smashed into the outer core coming inwards at around 40,000 miles per second. What is certain is that the star exploded, hurling elements into space, and leaving behind only 1.4 solar masses as a neutron core (a neutron star – incredibly dense, and only a miniscule remnant of the original star volume).

The non-scientist may view the above density with a sceptical eye; mutter darkly "mad scientists". But almost all the space that an atom occupies is due to the volume taken up by the electrons that surround the nucleus. The electrons elaborate the miniscule nucleus into an atom that is relatively vast. But, after gravitational collapse, the negatively charged electrons have combined with the positively charged protons to give neutrons. What remains is a core consisting simply of neutrons. Since nuclei have around 1/60,000th of the radii of atoms, nuclear volumes are around $(1/60,000)^3$ of those of atoms. Closely packed nuclear cores have densities around 10^{14} (100 thousand billion) times larger than sugar cubes. In neutron stars, fluffy atoms have been completely replaced by incredibly dense nuclear cores. It is analogous to a giant athletics arena being crushed down to the size of a 4cm ball bearing at its core. Realities outdo the fantasies of blockbuster Hollywood movies.

The discovery of neutron stars – a Nobel Prize winning achievement – is an example of the surprising leaps that can occur in science. In 1967, Jocelyn Bell and Anthony Hewish, scouring the heavens for radio-frequency signals at Cambridge University, discovered regular radio-pulses from what was at first known simply as a pulsar. Some wondered: "Were these signals transmitted by little green men?" for the frequencies (in the range of seconds to milliseconds) were initially difficult to understand. In fact, they were caused by a rotating neutron star – the residue from a supernova. The pulses occur at the same rate as the rotation of the neutron star, and the frequencies are, considering the large masses of the objects, remarkably high. The much larger stars from which neutron stars are formed rotate more and more quickly as their radius contracts. The tricks of neutron stars preceded those of the choreographers of ballet.

Cataclysmic star deaths can radiate more energy in 10 seconds than the Sun will radiate in its entire 10 billion year lifetime[6] (it still has several billion years of life in front of it). Strikingly, the supernova observed on Earth in 1987 actually occurred 160,000 years ago since the light from it took this time to reach Earth. In this time, the light travelled the equivalent of a succession of a billion journeys each of a billion miles.

But, most importantly for the journey of matter, it is stars that have provided the diversity of elements of which we are made. Our tiniest bits and pieces were in the ultimate cauldron of the "Big Bang", but these were then forged into larger units in the somewhat less mighty fires of stars. Given the rules and cauldrons that determine Earth's bits and pieces, we begin to get the feeling that our moulding, and that of our fellow organisms, will produce highly constrained beasts. And the rules obeyed within atoms add increasing strangeness to the story.

The Rules Obeyed within Atoms

During the course of evolution, some molecules were exploited in the birth of biology. At a later date, more complex molecules were selected by the constraints of biology itself. These latter molecules will be central to our story, for they determine both our forms and functions. They are the nuts, bolts, pistons, shafts and fuels of the machines of biology – very much our "makers". So, we cannot claim to understand ourselves without some knowledge of molecules, what they do and how they do it. But to know something about molecules requires a minimal understanding of the atoms from which they come. Fortunately, atoms are simple – no more than miniscule spheres. And the rules that determine their structures are also simple.

Elements – carbon, oxygen, iron, etc – are materials made up of only one kind of atom. The number of electrons in the atom – equal to the number of protons in its nucleus so that there is no net charge – determines the properties of the element. In a piece of revolutionary thinking, the Russian chemist Dmitri Mendeleev (born Tobolsk, Siberia in 1834, died 1907) noted that the properties of some elements are remarkably similar. For example, the elements helium and neon – of lamp fame – are very similar in their properties; both very stable gases. Yet helium has 2 electrons and 2 protons; whereas neon has 10 electrons and 10 protons. Similarly, lithium and sodium have very similar properties – both very reactive metals. Yet lithium has 3 electrons and 3 protons; whereas sodium has 11 electrons and 11 protons. Strikingly, for each of the similar pairs, the larger element has 8 more electrons (or protons) than the smaller one. Dmitri was so confident in his conclusions regarding these repeated properties – this "periodicity" – that he predicted the existence and properties of new elements. And, satisfying the requirement of first-class science, his predictions were frequently correct.

What is magic about the differences of 8? Dmitri could not contribute to an understanding of this problem for when, towards the end of his life, he was told about electrons, he was not a believer. But the answer did indeed lie in numbers of electrons – and balancing numbers of protons. In a two-dimensional "athletics arena" model of an atom (with the tiny nucleus at the centre of the arena but too small to be visible on the represented scale, Figure 1-1), the "fluffy" electrons are constrained in broad "running lanes".

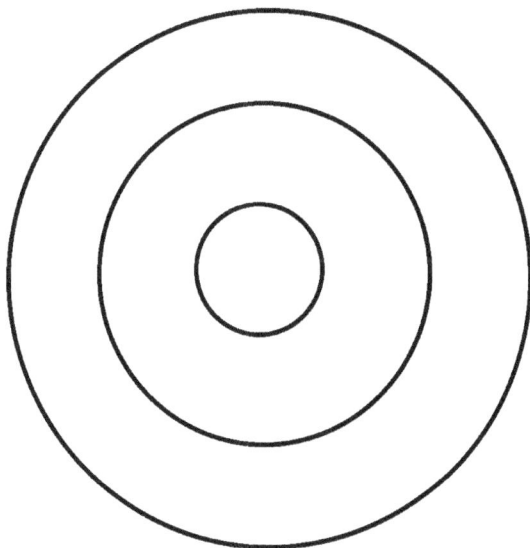

Figure 1-1. The "athletics arena" model of an atom: the innermost 3 "running lanes" for the electrons are shown as concentric circles.

Not only has the maximum number of electrons that any track can contain been established by experiment but it is also derivable from a simple mathematical relationship. If the number of the track is n, then the maximum number of electrons that it can contain is $2n^2$. Thus, the track nearest to the nucleus (n = 1) holds up to 2 electrons, the second up to 8, the third up to 18, the fourth up to 32, etc. The stability of atomic nuclei limits the number of elements to about 100, and the manner in which the tracks are occupied as they are successively filled makes no more than a total of 7 tracks relevant in the real world.

Note that the magic difference noted above for the simplest of Mendeleev's periodicities was 8 – the maximum number of electrons that can be accommodated in the second track. But *why* is it magic? The negatively charged electrons are attracted to the positively charged nucleus at the centre of the atom. So, as we progress along the first part of the Periodic Table, the electrons successively fill the first track with 2, and then the second with 8. Then they start to fill the third track, with occupancies of 8 (or finally the maximum of 18 for this track) being particularly stable. In general, partially occupied tracks are less energetically favourable than fully occupied tracks.

So helium (first track full with 2 electrons) and neon (first and second tracks full with 2 and 8 electrons) are both stable; and with similar properties, characterised by the message "both have similar track arrangements". Turn now to the pair of lithium and sodium – both unstable and similar metals. Lithium has the "successive track" arrangement 2, 1 – first track fully occupied, second track with one electron only. Sodium has the arrangement 2, 8, 1 – first and second tracks fully occupied, but third track with one electron only. Thus, lithium has similar properties to sodium. Additionally, both metals are similarly unstable because each has a single ("lonely") electron in its outermost occupied track, the energy of which is relatively high (further out from the positively charged nucleus than the others). Thus, the repetitions found in the properties of the elements that give rise to the famous Periodic Table are a consequence of them having the same number of electrons in the outermost occupied track.

Through the filling of the tracks 1 and 2 [to give a total of 10 elements (8 + 2) in all], there become available hydrogen, oxygen, nitrogen and carbon to use as the key components of organisms. Through the additional filling of track 3, oceans can be saline. With the additional filling of tracks 4 and 5, limestone cliffs and bones became possible; iron came about; and the Bronze Age became possible. With the filling of track 6, gold to hoard, and lead for the plumbing of the Roman Empire. With the additional filling of track 7, uranium for nuclear power. Many of the 100 elements have been exploited.

The ideas of electron tracks, their size and diffuse nature, and the maximum numbers of electrons that can occupy any particular track, are due to several scientists. Prominent among these are Niels Bohr, Erwin Schrödinger, Paul Dirac and Werner Heisenberg.[7] Niels Bohr was born in Copenhagen in 1885 into an academic family, and extended the Rutherford model of electrons as orbiting the nucleus (the reasons for the demise of this model are covered in chapter 3). For his work on the structures of atoms, he was awarded the Nobel Prize in Physics in 1922.

Erwin Schrödinger was born two years after Bohr, in Vienna, and served in the First World War as an artillery officer. Following his greatest discovery (the Schrödinger wave equation) in 1926, his work led to a very successful model of the electrons behaving as waves. In 1927, Schrödinger moved to Berlin as the successor to Max Planck (a key player in the development of quantum theory). With the coming to power of the National Socialist Party in 1933, he decided he could not continue to live in Germany. Following numerous appointments at different locations, he eventually became Director of the School for Theoretical Physics in Dublin and remained there until his retirement in 1955. Schrödinger had a more colourful reputation than most scientists with oft-repeated tales of his "ménage à trois" household.

Paul Dirac arrived on the scene somewhat later, being born in 1902, in Bristol, England, of a Swiss father and an English mother. In 1932, he became Lucasian Professor of Mathematics at Cambridge – following in this respect the footsteps of Isaac Newton, and anticipating those of the theoretical cosmologist Stephen Hawking. He shared the Nobel Prize in Physics with Schrödinger in 1933. Stories abound of Dirac's logical approach to life, and his economy with words. It is told that, shortly after he received his Nobel Prize, he was asked by a student: "May I study for a PhD under your supervision?" "No," replied Dirac. After a long hesitation, the student ventured, "Why not?" "Because I seem to be doing quite well on my own, thank you," replied Dirac. On another occasion, when asked, "How did you find the Dirac equation?" he responded, "I found it beautiful."

So Nature has the possibility to build from about 100 building bricks – the elements, the atoms of which are built up according to remarkably simple rules. As will be seen later, Nature uses remarkably few of these bricks to build most of our bodies.

The Birth of our Makers

How does a newness come into the world? How is it born?

– Salman Rushdie in 'The Satanic Verses'

Newness comes into the world – or, more widely, the Universe – through the making of new combinations and interactions. And so it was when a new combination was born in a cooling Universe – molecules through the combination of atoms. Do not neglect molecules for, although religious leaders have told us for millennia that gods are our makers, in truth our makers are molecules. And what wonders they perform.

All molecules, and compounds, are simply combinations of atoms. Atoms, by virtue of their spherical symmetry, are rather boring and simple structures. But their combination in molecules allows the generation of an almost limitless variety of shapes. The devices of biology, and hence organisms, become possible. The length of the smallest molecules is about a hundred times less than a millionth of a centimetre. However, where their constituent atoms are joined together in long strings – say some stretches of DNA – the molecules become proportionately longer.

In a cooling Universe, at what temperature, and roughly when, did the first molecules appear? Once the temperature in parts of the Universe had dropped to around 10,000°C, hydrogen atoms could join together in pairs to give molecules of hydrogen (H-H, or H_2). This simple and relatively stable molecule first appeared in the Universe about 100 million years after the "Big Bang", 13.6 billion years ago – around the time that the first stars also formed (Table 1-1). At a much later time – roughly 10 billion years after the "Big Bang", and therefore about 4 billon years ago – when the temperature of a part of the Universe (Earth) had dropped to about 100°C – molecules that were much more delicate in their make-up could both form and persist. The building blocks that enabled biology – our makers – began to occupy the stage.

Why do atoms combine to give molecules? They do so because the combination is the more stable system at the prevailing temperature. The 100 different kinds of atoms can join together in so many ways that there is an unimaginably large pool of possible molecules and assemblies of them. So, making rocks, water, air, concrete, steel, custard pies, eels, butterflies, humans and hundreds of millions of other things is not a problem. But these items are only around in a given environment because that environment is sufficiently cool to avoid their disruption. Put a custard pie into the interior of the Sun and its constituents almost instantaneously revert to nuclear cores in a sea of electrons.

So, an understanding of the origins and building of humans requires knowledge of the rules that molecules must obey. Experiments show that, from the atoms H, C, N and O, simple molecules that are formed include the gas methane (always CH_4), the gas ammonia (always NH_3) and water (always H_2O). Why do these simple rules apply?

As already noted, atoms with tracks partially filled with electrons are less stable than atoms with fully filled tracks – neon, a stable element since it is an atom with a full second track containing 8 electrons. But the atoms of carbon, nitrogen and oxygen that precede neon in the Periodic Table are less stable because this same track contains, respectively, only 4, 5 and 6 electrons. To attain systems of greater stability, atoms join together and *share* their electrons to build full tracks.

The hydrogen atom possesses only 1 electron and, in order to satisfy the electromagnetic force by getting as near as it can to the positively charged nucleus, this electron occupies the first track. The oxygen atom possesses a first track filled with 2 electrons, but only 6 electrons in its second track (left-hand side of Figure 1-2, with first filled track not shown). Since a first track requires 2 electrons for a stable arrangement, and a second track requires 8 electrons for a stable arrangement, full tracks can only be attained by sharing electrons – achievable when the hydrogen and oxygen atoms get together in the ratio of two to one. In the aggregate H-O-H, out of its 6 second-track electrons, the oxygen atom shares 2 of them with the 2 hydrogen atoms. So each of the 2 hydrogen atoms acquires 1 electron, which is added it to its existing 1 electron. The hydrogen atoms thereby attain, in H_2O, the 2 electrons required for a fully occupied and stable first track (right-hand side of Figure 1-2).

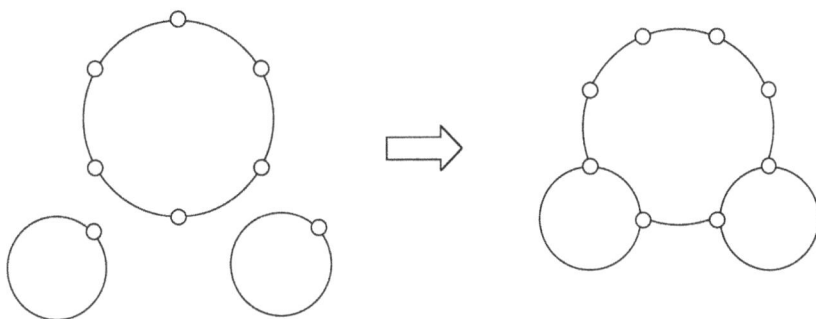

Figure 1-2. The rule for making water – our most abundant component. (The "fluffy" electrons are conveniently shown as tiny circles.)

Conversely, and logically connected, the 2 hydrogen atoms each share the 1 electron that they originally possessed in their first track with the oxygen atom. Thereby, the second track of the oxygen atom increases the occupancy of its second track from 6 in the isolated atom to the stable full occupancy of 8 in H_2O. In a water molecule written out as H-O-H, the lines connecting the atoms represent the attractions that join them together through the sharing of electrons. In the parlance of the chemist, they represent "bonds". Through the device of sharing, the atoms became a more stable combination in the molecule H_2O. Once the cohesive glue provided by this sharing was greater than the tendency of the constituent atoms of H and O to bounce apart because of their agitation through motion then water became a constituent of the Universe. A key ingredient exploited in the future evolution of biology became available.

In addition to hydrogen, there are three other elements of central importance in biology (C, N and O). They have respectively 4, 5 and 6 electrons in their outer track. So the sharing principle requires that when they combine with hydrogen atoms, they will acquire a number of hydrogen atoms so that the total number of electrons in the track becomes 8. Hence, the resulting molecules are CH_4, NH_3 and H_2O. In making connections to other atoms, the rule always works. It accounts in a beautiful and unifying way for the constitutions of molecules: whether it is in the DNA that codes for us, the proteins that allow our function, or the vitamins that sustain us, carbon makes four connections, nitrogen makes three, and oxygen makes two. Only some 200 years since the first molecular structures were determined, we now have methods and rules to determine the molecular structures of almost all our makers. In conjunction with other cardinal rules, the "sharing rule" determines the nature of the world around us, and of ourselves. It constitutes one of the great scientific revolutions of the twentieth century – surely worth a Nobel Prize? Well, no actually.

The concept of electron sharing to produce bonds was due to Gilbert Newton Lewis. "GN", as he was commonly known, was born in Massachusetts in 1875. He was appointed the chairman of the Department of Chemistry and the Dean of the College of Chemistry at UC Berkeley in 1912, positions he held until his sudden death in his laboratory on 23 March 1946. He did not let his administrative duties interfere with his chemistry but ran the department from his laboratory, making decisions and dictating letters while conducting experiments. Although many realised the supreme importance of GN's contributions to understanding the structures of our makers, as reflected in his nomination for the Nobel Prize in Chemistry on numerous occasions between 1922 and 1935, he never got it. The system failed him.

But systems of particles behave more reliably. When the temperature of parts of the Universe dropped sufficiently, then colliding molecules should, under the further influence of the electromagnetic force, aggregate to give more complex forms of matter. And aggregate they did – with particularly remarkable consequences in the unusual environment called Earth, which must next grab our attention.

2

From Stars to Earth

The journey of matter from the fire of the Big Bang, into the frying pan of stars, and the explosion of the stellar frying pan – hurling its content into space – has been recounted. Remarkable as these journeys are, the most astonishing transformation was yet to come. The products from the stars spontaneously aggregated to become – quite literally – a living room (Earth), with the forms of life implicit in that description.

Birth of the Solar System

When a star ends its life in the explosion described as a supernova, the heavy elements that were synthesised inside it (with the accompanying residual light elements of hydrogen and helium) are not condemned to wander in space forever. Rather, in a process analogous to the formation of the first stars, the dust particles may recondense under the influence of their mutual gravitational attraction. About 4.55 billion years ago (about 9 billion years after the "Big Bang"), one system of inter-stellar dust began such a gravitational collapse. The energy released by this collapse caused the material near the centre to become an extremely hot core. A new cycle of nuclear fusions could begin in this region; the Sun was born. But why did the whole system of dust not collapse into the Sun? Why the planets?

Although it is not known whether the Universe rotates as a whole, spiral galaxies are the Catherine wheels of the cosmos. Galaxies do rotate. This occurrence of local rotational motions within the Universe provides a centrifugal force – the force that stretches the arms of the hammer thrower. It opposes gravitational collapse. Like the hammer thrower in mid-performance, the stellar dust and gas that led to the solar system was also rotating. So the material more distant from the centre did not condense into the incipient Sun. It condensed into the orbiting planets. As necessary for formation from a rotating disc, the planets orbit in the same direction round the Sun.

The emanations from the explosion of a stellar frying pan had, guided by universal rules, given birth to the solar system. It remained a mystery to Newton why the planets were "set up" with their orbits almost in the same plane.[1] It is a natural consequence of their formation from a spinning disc of stellar dust and gas. The loose skirt of a ballet star, spreading out in a near to horizontal plane when a rapid pirouette is performed, does not dance alone. It follows the same rule as the solar system.

Supersonic Speed and Rotation of the Living Room

Earth lies about 93.5 million miles from the Sun and, by definition, takes one year to orbit round it. Our living room is travelling in a roughly circular path at a speed of 67,000 miles per hour – about one hundred times the speed of sound.

Simple physics dictates that planets formed by aggregation from a rotating disc will spin on their own axes in the same direction as the disc itself – anti-clockwise when viewed from the side of the disc from which our North Pole would be seen. So the Sun rises in the east and sets in the west. Such is the rate of the Earth's rotation on its own axis that our day has a length of 24 hours. So, besides speeding along at 67,000 miles per hour, we additionally rotate at 1,000 miles per hour.

With regard to the direction of axial rotation, Uranus and Venus are renegades: sunrise in the west. The reversals are probably due to giant impacts – historical commonplaces evident from the craters on the surface of the air-less Moon. But from whence the Moon? An impact of a giant meteor upon the Earth produced recoiling debris that later condensed to become the Moon; an impact with giant consequences.

Impacts of Rotation and Impact

The same impact as led to the Moon probably caused the Earth to tilt its axis of rotation. For, contrary to the expectation that the axis of rotation of the Earth would be at a right angle to the plane in which the solar system rotates as a whole, it is tilted by about 23°. The effect of the tilt is to alternately project, once every six months, either the northern or southern hemispheres more directly towards solar radiation.

The rotation of the stellar material from which we came, and a giant impact upon it, has impacts of other kinds: trivially, no Thomas More as *A Man for all Seasons*; no *Four Seasons* to arise from the mind of Vivaldi. More substantially, there would be no tides, no deciduous trees, no migrating swallows and no hibernation. Without axial rotation, no bedtime after dusk and no alarm clock at sunrise. Forms and behaviours were fashioned by local angular momentum and chance collision.

This knowledge was not accrued easily; nor without pain. That the Sun was "the centre of the heavens", with the planets circulating round it, was proposed by Nicolaus Copernicus and published in the year of his death (1543). Almost a century later, Galileo Galilei (1564-1642) built a telescope, and provided further support for the proposal of Copernicus. He was in conflict with the Roman Catholic Church by 1633, at a time when the church dogma stated, "Propositions to be forbidden: that the Sun is immovable at the centre of the heaven; that the earth is not at the centre of the heaven."

Galileo's story is a classic case of the truth that can come from observation and experiment, and how it may be opposed by a dogma based on ignorance. For his heresy, Galileo's punishment – house arrest – was mild. His contemporary Giordano Bruno was less fortunate: he was burned at the stake in Rome for the same heresy.

A Venerable Living Room

The above pioneers would have been astonished by the great age – 4.55 billion years – of the system they studied. How is such an enormous age justified?

The history of estimates of the age of the Earth is instructive. Quantum leaps occurred in them as humanity progressed from ignorance to knowledge based on experiment. The traditional religious dogma spoke of an Earth created only a few thousand years ago. But the Scotsman James Hutton, often regarded as the father of geology, in 1787 and 1788 noted rock structures that were not only formed in layers, but in layers that had been twisted at large angles with respect to each other.[2] These simple observations led him to question the at-a-stroke formation of a relatively youthful body.

The great physicist Lord Kelvin knew big problems when he saw them. He, who had wrongly wrestled with the age of the Sun, also wrestled with the age of the Earth. In his estimation, he used the rate of loss of heat through rock, and made an assumption regarding the temperature at the core. From the current heat flux at the surface, he came up with an age for the Earth of 20-400 million years (most probable value, 100 million). But Ernest Rutherford put the fat in the fire – or least new coals into it.

From his knowledge of radioactivity, he inferred that this process would continuously produce heat within the core region and would invalidate Kelvin's estimate.

The need was for something more reliable. That something was the rate of radioactive decay. The nuclei of some atoms are, rather unusually, unstable. They decay to give another nucleus at a characteristic rate. By this means, one element is transformed into another. This transformation is associated with radioactivity [discovered by Antoine Henri Becquerel (1852-1908)], because the transformation results in part of the nucleus being expelled as particles of high energy.

The understanding of radioactivity was greatly extended by the researches of Pierre Curie (1859-1906) and Marie Curie (born Maria Sklodowska, in Warsaw in 1867, died France in 1934). The early death of Pierre was caused by a traffic accident in Paris in 1906, shortly after Becquerel and the husband and wife team of the Curies had shared the 1903 Nobel Prize in Physics. Marie, famous for her quiet and unassuming manner, received a second Nobel Prize, this time for Chemistry, in 1911. Her death was caused by leukaemia, perhaps due to her exposure to the energetic particles (radiation) expelled during radioactive decay. So severe was her exposure that her documents are characterised by high levels of radioactivity. I was unaware of this when, in 1966, I perused without fear a copy of her thesis – owned by my colleague Franz Sondheimer – given and dedicated "à Monsieur Rutherford".

In radioactive decay, the rates of decay are characteristic for each type of element that undergoes the process. These rates (measured in terms of the time needed for half of the starting element to be converted into the product element) span the useful range from a few years to billions of years. The age of a rock is defined as the time that has elapsed since it was formed from the molten state. Usefully, rocks are solid structures from which any gases formed within them, by radioactive decay, cannot escape. So when, in the interior of rocks, radioactive potassium decays into argon gas, the present-day determination of the argon/ radioactive potassium ratio allows the determination of the time that has elapsed since the rocks were formed from the molten state.

The decays of radioactive potassium to argon, and of radioactive uranium into lead are very slow; so slow, they can be used to measure the time of formation of rocks of the order of a billion years old. The method of dating through radioactive decay is an absolute one. Securely, the oldest rocks on Earth are in the region of 4 billion years old. The age of the solar system (4.55 billion years) comes from dating of the rocks of meteors (themselves part of the solar system) that have fallen upon Earth. So, the values established from rates of radioactive decay are about a million times greater than the one based upon religious tradition, and about a hundred times larger than the one (Kelvin's) based upon a mid-nineteenth century knowledge of physics.

It is a measure of human susceptibility to received wisdoms that are contrary to experimental evidence that a considerable fraction of the current US population believes that the Earth is less than 10,000 years old. These beliefs underestimate the experimental value by a factor of about 400,000. The belief is analogous to insisting, despite the evidence to the contrary, that the human lifespan is about one hour. Methuselah dying at 969 is more readily swallowed by some than is the reliability of radioactive decay.

The Continents – Floaters and Drifters

Given a lot of time, there can be much movement. The flowing nature of the hot material that underlies the cooler and solid surface of the Earth allows the continents to migrate ("continental drift"); the whole of the Earth's crust is in motion. In the Devonian period (covering a period 400 million years ago), India was well south of the equator. It then rushed north and ended up, 40 million years ago, bumping against the main Asian landmass. Hence the Himalayas.

The dating of such remote events is a thing of beauty. What has happened in the Atlantic Ocean during the last 100 million years makes the point. During this period, there has existed, near its middle, a fissure running in a north/south direction. Molten rock, from the interior of the earth, oozes through the fissure and then solidifies as rock at the ocean bottom; or, less commonly, at the top of the ocean, as evidenced by the existence of Iceland. Propelled by the continuing ooze, this rock then moves either to the west or to the east. The continents closest to this rock are effectively floating upon, and moved by, it. Those of North and South America move to the west, and those of Africa and Europe move east. The Atlantic is getting wider.

During this time, the Earth's magnetic field reversed in a relatively regular manner. Reversals were slow – they took a few tens of thousands, or even up to hundreds of thousands of years. Some of the molecules in the initially treacle-like rocks of the ooze behave like tiny compass needles. They therefore orient according to the two possible opposing directions of the Earth's magnetic field at the time when the rock appeared from the bowels of the Earth. When the rock solidifies, these orientations are retained. Therefore, by counting the number of reversals of the magnetic particles as a function of distance from the central fissure, the positions of the continents as a function of time can be estimated.

The reversals of the Earth's magnetic field are relatively slow, but the drift of the continents is slower. South America has been chugging away from Africa for somewhat in excess of 100 million years. Although the continents drift at a very slow pace (about the rate of growth of a finger nail), great distances are covered over periods of hundreds of millions of years.

When the belts of moving rock reach the floating continents of Africa and South America, they are then subsumed beneath the continents and pass back to the interior of the Earth. But the surface rocks of the continents stay afloat like crust at the top of boiling milk, and hence are much older than the continuously recycled rocks of the oceans' beds. Continental drift – plate tectonics – was accepted only gradually during the twentieth century for the truth was too surprising.

Once the times of the migrations of these enormous landmasses was known, it was possible to establish how long species had been isolated on their respective "life rafts"; when and where different kinds of flora and fauna evolved. There was still a connection between Australia and South America 55 million years ago, so the fact that marsupials appear on such widely separated southern continents was no longer perplexing. Evolution had discovered a different dodge for parental care in the northern and southern hemispheres. The boats on which organisms sailed led to adventures more remarkable than the myths of Odysseus.

The Earth's interior may be in molten turmoil, but the crusts are cooler; and cooler crusts allowed molecular aggregates to form. The stuff from which organisms would evolve appeared.

Aggregates of Molecules

The mob has many heads, but no brains.

– English proverb

The "goldilocks" temperature zone for making molecular mobs and webs is around 0-100°C. The surface of the Earth is a place that fulfils this requirement; whether "a special place" is a philosophical point. For, given variation among the around 10^{21} or more objects in the cosmos, much is possible, and even probable.

A human being, or any animal of similar size, is constituted from around one thousand trillion trillions (10^{27}) of molecules. These molecules interact with each other in webs of Byzantine complexity. Life is not only dependent on the possibility of forming such inter-molecular webs, but also requires the possibility to reorganise – make and break – them. Yet, in popular accounts of the why's and how's of life, molecules get short shrift. But from inanimate webs of molecules came life itself. The path to life, and its processes, required another step up the binding hierarchy – from molecules to their aggregates.

When the temperature of parts of the Universe had dropped to less than around 1,000°C, some molecules could associate with themselves, or with other molecules. At less than 100°C, isolated water molecules could join together to give a collection of countless trillions of water molecules – bulk water. In all aggregates, properties not readily predicted from the constituent parts emerge. The mobs of city riots take on properties not found in, or predictable from, the behaviour of individuals. Liquid water similarly confuses. It exhibits properties not yet predictable from the behaviour of an isolated water molecule.

Why do molecular mobs emerge in cool environments? They emerge because collections of large numbers of + and – charges can produce weak attachments. Imagine the large attraction between one positive charge and one negative charge at the following distance:

$$+ \quad -$$

Next, imagine the attraction between these two charges at the same distance after an extra negative charge is "hidden" immediately behind the positive charge:

$$-+ \quad -$$

The negative charge to the right is now attracted to the - + combination less than to an isolated + because it also experiences repulsion due to the newly added negative charge. From the view of the original negative charge, the - + combination acts as a partial (fractional) positive charge. Due to this kind of effect, in water molecules, the oxygen acts as though *slightly* (partially) negatively charged ($\delta-$), and the hydrogen atoms as though *slightly* positively charged ($\delta+$). This makes for a weak attraction, the hydrogen bond (...), between them: adjacent molecules join as H_2O... H_2O...H_2O...H_2O..., elaborated into three dimensions. Without the hydrogen bond, no ice, no liquid water. But, much more – for all life depends upon such weak attractions: it is through their courtesy that the key molecular aggregates of biology – the double helix, and folded and aggregated proteins – exist.

The importance of the hydrogen bond, both in the molecules of biology and elsewhere, was made clear by the great American chemist

Linus Pauling,[3] who died in 1994 at the age of 93 in California. He was the only person ever to receive two unshared Nobel Prizes – one for Chemistry in 1954, and one for Peace in 1962. He also defined the nature of the most important structural motifs that are found in proteins (chapter 4), and was the first to demonstrate the molecular basis of disease. Outspoken against atmospheric nuclear testing, he was denied a passport to travel to conferences during the infamous McCarthy era in the US in the early 1950s. He was a scientific giant.

Why Land, Sea and Air?

Attractive forces between molecules can be very weak, weak or even moderately strong; hence – respectively – gases, liquids and solids. For this reason, it was eventually possible to evolve birds that would fly in air; fish that would swim in sea; ants and humans that would walk on land. For this reason, organisms could evolve to breathe air; to pump blood; to stand on bones.

If the forces between molecules are towards their lower limit then, at the temperatures pertaining at the surface of the Earth, the molecules separate – they exist as a gas (eg the nitrogen of the atmosphere). If a little stronger, the molecules have enough cohesion just to hold together, as in a liquid. Ice and water illustrate the key differences between solids and liquids. If the temperature is below 0°C, the molecules align like soldiers in fixed ranks on parade: a solid (ice). In the range 0-100°C, the water molecules are more agitated; they stick together only loosely – again like rows of soldiers on parade, but where the "bonding" between adjacent soldiers is sufficiently loose that they are constantly changing places within their ranks. This "changing places" among adjacent molecules accounts for the flowing of liquids – they take up the shape of the vessel containing them.

Opposite and adjacent electrical charges that appear *partial* (preceding section) typically give rise to liquids and to solids that are readily melted. The water/ice system is a classic case for the inter-molecular attractions are rather weak. In contrast, arrays where adjacent atoms and molecules possess alternating positive and negative *full* charges make less tractable solids – as in rocks, earth and table salt. But how do adjacent full charges arise?

How electron *sharing* can produce stable tracks of electrons, and so give rise to stable molecules, was illustrated in Figure 1-2. But there is a second way to produce stable tracks of electrons – to *transfer* an electron from one atom to another. Such transfer is favoured if the donor atom has only one electron in its outer track to start with, and the acceptor atom requires one electron to obtain a stable configuration of its outer track. The transfer from donor to acceptor is a "win-win" situation, for both participants end up with tracks occupied to an extent that gives great stability (Figure 2-1).

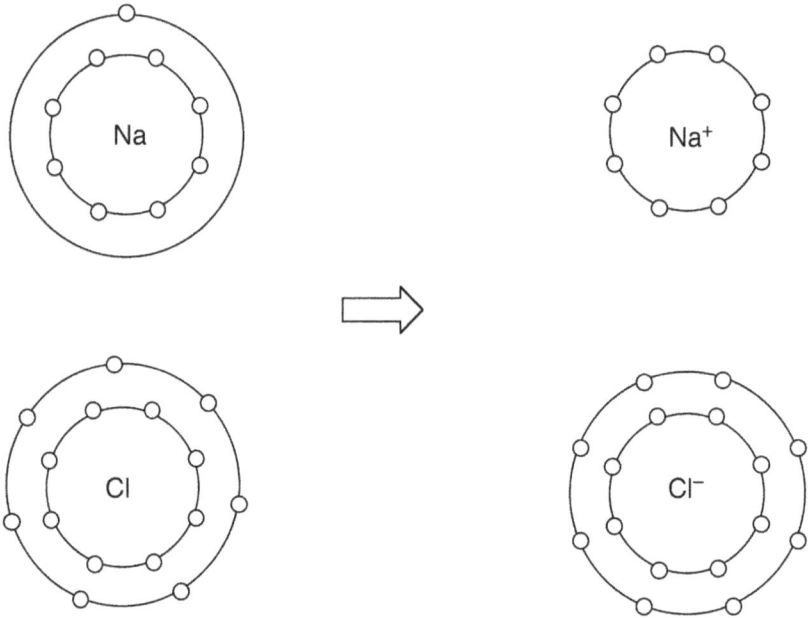

Figure 2-1. Unstable atoms of sodium (Na) and chlorine (Cl) transfer an electron from one to the other, and so become a stable combination of sodium ion (Na^+) and a chloride ion (Cl^-), each with 8 electrons in their outer tracks.

In understanding the contents of the Earth, these changes are not trivial, in this specific case telling us why it is not easy to possess sodium or chlorine atoms, but that the oceans can collectively contain around 5,000 trillion kilograms of stable table salt. Through the transfer of a negatively charged electron from one atom to another, the acceptor acquires a charge -1; the donor a charge +1. Consequently, there is a remarkably strong electrical attraction between adjacent receiver and donor particles. Electron transfer between atoms strongly constrains the products to remain as solid aggregates, for extended -+-+-+ arrays in three dimensions are strongly constrained (exceptionally, water is in some cases able to break down these aggregates). At a stroke, the commonly intractable solid structures of rock, salts and earth, are comprehensible. In attempts to understand the Universe, the concept of electron transfer represents, like electron sharing, another stunning advance. There is a sudden enlightenment.

Whilst the products produced by electron sharing are called molecules, those produced by electron transfer are commonly known as compounds. Because of these two different modes of attaining stability, the Earth and its occupants are the way they are. In the living world, mobility commonly serves survival. Therefore, bodies make major use of the products formed from atoms by electron sharing, and subsequent weak aggregation, to give "bendy" products. By contrast, mountains, their rigid minerals and rock-like bones, are mainly the products of electron transfer. The rules of stability are such that there is a fine balance between transfer and sharing. Which provokes a philosophical question:

Is it an Improbable Universe?

Three chapters have covered a lot of evolutionary history. We have a widely accepted model of what the Universe was like around one microsecond after the "Big Bang". It was remarkably homogeneous – relatively smooth, rather as we envisage the structure of gaseous water in a boiling kettle to be smooth, with the tiny particles within it all colliding with each other. It was not lumpy.

But neither was it perfectly smooth. Like the gas in the kettle – which condenses to form water droplets after expansion through the spout – it contained irregularities. Some might say "fortunately for us", since without these it would have remained a featureless gas; none of the small irregularities that could have resulted in local rotations; none of the small irregularities necessary to nucleate stars and galaxies. In short, nothing of interest would have happened.

If, on the other hand, the very early Universe had been *very* heterogeneous ("lumpy") then at a relatively short time later (and long before life would have had a chance to evolve), it would have condensed into relatively few massive lumps. So, we must live with a rather remarkable fact – the degree of smoothness was just right for the later emergence of a Universe rich in structure.

This feature is one of the so-called "fine-tuning" arguments.[1,4,5] For example, the conditions of the "Big Bang" were such as to control the ratio of two very different kinds of particles. There were *photons* – the particles that have no mass and travel at the speed of light; their transmission corresponds to radiation (radio waves, light, X-rays, etc). There were also *baryons* – those particles associated with matter (eg the protons and neutrons of Table 1-1). The ratio of photons to baryons was approximately 10^9 (a billion to one). A billion photons for every baryon is a very large ratio but, without the baryons, the Universe would be featureless – without atoms, stars and planets.

These two above properties of the Universe, in conjunction with the previously described balance of its initial kinetic energy and eventual potential energy, seem to be incredibly improbable. However, there may be as yet undiscovered fundamental laws that demand these three consequences in the "Big Bang". These "fine-tunings" involved in the evolution of the Universe, to which can be added the fine balance between electron sharing and transfer, leads to the manifestly true "Weak Anthropic Cosmological Principle" – that we are here only because the Universe has a number of remarkably delicate balances. The extension of this argument to the idea that the Universe was uniquely *designed* for us (one view of the "Strong Anthropic Principle" considered by Barrow and Tipler[6]) is however a Big Jump – as evidenced in later chapters.

Matter still has much further to go, and quantum mechanics and thermodynamics are crucial in understanding its journey towards life. These subjects deter many. But their basic concepts, to which we next turn, are simple.

3

The Rules for Spontaneous Order

Our story now progresses towards "How come organisms?" On the basis of what has gone before, the reader must feel scepticism that they can appear without magic. Although it really came as no surprise that, in a cooling Universe, bits of matter increasingly clumped together with the passage of time (for water vapour does condense on leaving the kettle), when wisteria, worms and wombats appear, their organised forms are not generated through cooling. The principles that tell what is going on emerged from the latter part of the eighteenth into the twentieth century. They remove the necessity for the earlier used woolly descriptions that atoms combine to give molecules, and molecules aggregate to give liquids and solids, because the larger aggregates are "more stable" or "energetically more favourable" at a *specified* temperature. They allow wisteria, worms and wombats to spontaneously appear. As related here, the story is not chronological.

Small Things are Queer

Quantum theory is crucial in making successful predictions when dealing with the very small – things that, even with the help of the most powerful light microscope, we cannot see. But this is of little relevance to the behaviour of objects that we see – which is probably why the coming of quantum theory took so long. One of its conclusions is that, in observing the behaviour of particles of small mass (eg electrons), the method of observing (eg light) may itself modify the behaviour. Hence, we can never know precisely how such miniscule systems behave. There is an in-built degree of uncertainty.

The degree of uncertainty was quantified in the famous "uncertainty principle" of Werner Heisenberg (born Würtzburg, Germany in 1901, died Munich 1976). The basic message is that the smaller the momentum (the mass times the velocity of the particle) of the system that you wish to pin down, the larger will be the uncertainty in your measurement.

In 1932, he was awarded the Nobel Prize in Physics for "the creation of quantum mechanics". During the Second World War, he headed the unsuccessful German nuclear weapons' project. As a consequence, he was arrested at its end and interned in England; but returned to Germany in 1946 as director of the Max Planck Institute for Physics and Astrophysics at Göttingen.

The success of quantum mechanics indicates that the Universe is, at small scales, a place that it is difficult for the human brain to intuit. Perhaps inevitably so, for that organ evolved to promote survival in macroscopic surroundings. It had no understanding of electrons and protons, for the survival of its owner had not required it. This realization prompted, from the evolutionary biologist J B S Haldane, the comment: "The Universe is not only queerer than we suppose, but queerer than we can suppose." When two atoms are held together in a molecule, it is not queer to envisage the molecule as a barbell, with the atoms at the end of the connecting rod. Not queer that experiments tell us that the barbell vibrates in and out along its length, for the atoms have thermal energy and refuse to be pinned down. But definitely queer that quantum mechanics tells us, and experiment agrees, that the vibrations of the molecular barbell can only occur at specific frequencies.

Energy Comes in Lumps

Energy comes in lumps; and why not, for we very happily accept that mass comes in lumps.

– Jacob Bronowski

In the first part of the twentieth century, other signs appeared which indicated that our "intuitive" view of the world could be misleading. The description of "light" posed a dilemma. Light as a "beam of photons" conveys an impression of bullets fired from a machine gun. But light – and, more generally, radiation – is also considered as waves. This leads to a long-standing dilemma for the student: "But light cannot be both a particle (photon) and a wave – it must be one or the other." But scientific models are not devised necessarily to be unique descriptors; rather, if fruitful, to be used. Both the particle and the wave models further an understanding of radiation, so both are in use.

But the student acolyte does have a point; for the acceptance of both the particle and the wave models is counter-intuitive. And there were worrying signs that the Newtonian world of "classical physics" was cracking up when applied to the very small. The atomic model of electrons circulating round nuclei was initially embraced because there was an appealing analogy to the circulation of the Moon around the Earth – the classical Newtonian world. In this model, the circulating electron must be accelerating. For, during the course of one half of a revolution, its velocity changes from whatever value it had in one direction to an equal velocity in the opposite direction. An accelerating electron is required to emit radiation, and – horror of horrors – such an emission does not occur. But if the electron was not in circular motion, why did it not simply fall into the nucleus (as a non-circulating Moon would fall to the Earth)? Newtonian mechanics was not applicable, and the "circulating-electrons model" was a pipedream.

Newton had seen a world of *big* objects, such as pendulums, and derived the Newtonian mechanics that govern their behaviour. When energy is added to a pendulum to give it a wider oscillation, there appears to be no restriction in the amount that is added. When a car is given energy, it can be added without a sudden jump from 10 to 20 miles per hour – the input of energy seems seamless. So it was natural that the physicists of the nineteenth century assumed that *any* particle could, irrespective of its size, possess any amount of energy. In other words, energy did not come in "lumps".

But this assumption was wrong – energy *always* comes in lumps – known as *quanta*. The electrons of atoms can have only certain specified energies in their diffuse tracks (or shells) around the tiny nucleus. It is as though satellites could only be put into orbit round the Earth at distances of, say, 2,000, 5,000 and 10,000 miles – clearly not true for massive satellites; but it is the case for electrons of miniscule mass within atoms, and they stay "up there" without orbiting. A photon of radiation "fired" at an atom can cause one of the electrons of the atom to jump from an inner to an outer track (or shell) – but only when the energy of the photon corresponds to the fixed energy difference of the electron on changing from the inner to the outer track. In sum, in the jump of an electron from an inner to an outer track, only specific "lumps" of energy can be absorbed.

Even in the case of a pendulum, the energy is added in lumps. It is just that the added quanta are so tiny compared to the total added energy that it *appears* energy can be added in any amount. Although these lumps, or packets, of energy are miniscule compared to the energy that we add to a clock pendulum, on their own tiny scales they vary enormously in size. And variation in size has deep consequences for what happens in the Universe.

Ludwig Boltzmann

To go straight to the deepest depth, I went for Hegel; what unclear thoughtless flow of words I was to find there! My unlucky star led me from Hegel to Schopenhauer... Even in Kant there were many things that I could grasp so little that given his general acuity of mind I almost suspected that he was pulling the reader's leg or was even an imposter.

– Ludwig Boltzmann [Source: D Flamm, Stud Hist Phil Sci 14: 257 1983)]

It must be splendid to command millions of people in great national ventures, to lead a hundred thousand to victory in battle. But it seems to me greater still to discover fundamental truths in a very modest room with very modest means – truths that will still be foundations of human knowledge when the memory of these battles is painstakingly preserved only in the archives of the historian.

– Ludwig Boltzmann

There is only one arrangement of an unbroken bar of chocolate. But once the bar is broken into (say) two identical squares, these squares (1 and 2) can be used to reassemble a complete bar in two different ways (2-1 or 1-2). A combination of bits is more probable than a single entity, as seen in the results of releasing a bull in a china shop. And so it is for energy. If a given quantity of energy can exist in one lump, or alternatively be in several smaller lumps, the latter state is more probable.

The great Austrian theoretical physicist Ludwig Boltzmann (1844-1906) deduced a quantitative relationship showing that the larger the number of ways (W) of distributing the energy in a system, the greater the probability that it will exist in that state. He described the probability of the state (S) as the *entropy* of the system. In a nutshell, a system with a large W is preferred over a system with a smaller W, and has a larger entropy S. The take-home message is that *change in the Universe occurs with an increase in entropy, for systems where the energy is contained in smaller packets exist with higher probability.*

Boltzmann's equation (S = k.lnW, where lnW represents the logarithm of W, and k is a constant) is carved on his tomb in Vienna. Like many other equations – those of Newton, Einstein, Maxwell, Schrödinger and Heisenberg – it derives from the efforts of mathematicians seeking to increase our understanding of physics, and hence of the Universe. The fact that these concise equations, born of human imagination, describe and predict the behaviour of energy and matter in the real world both humbles and amazes. Whether they represent eternal truths, or provisional laws, we can only guess.

Boltzmann's discovery puts him up there with Newton, Einstein and Darwin. But he is little trumpeted. He committed suicide in 1906, perhaps in part because his pioneering ideas were not initially widely accepted. But he was way ahead of them; his opponents were the ostriches of science. For entropy is arguably the most important concept necessary for an understanding of happenings in the Universe; and one of the least understood.

Small and Large Lumps of Energy

What determines whether the energy comes in small or large lumps? The mass-less photons transmit energy (eg as visible light) from the Sun to Earth. The photon particles also have, because of their additional wave qualities, a specific frequency – the number of waves per second. Those corresponding to light have frequencies in a relatively narrow band, differing in a characteristic way according to the colour: blue light with a photon vibration of close to a thousand trillion times per second (10^{15} s^{-1}), and red light with a photon vibration about half this value.

Stemming from the work of the German physicist Max Planck, we know that the higher the frequency of photons, the larger are their energy packets (their quanta); the energy packets of blue light larger than those of red light. Planck was awarded the Nobel Prize in Physics in 1918 for his fundamentally important discovery. It is related that, when on a walk with his son, he told him that he had made a discovery worthy of Newton.

Einstein carried the idea of Planck from the behaviour of radiation to the behaviour of the oscillations found in the world of mass. He applied it to the rates of vibrations – frequencies – of atoms within molecules. The same concept applied. For, in a comparison of the vibrations of two types of atom-atom bonds (atom-to-atom connections) in a molecule, the vibration with the lower frequency requires smaller packets of energy to bring it about. So, if small energy packets within a system make its existence more probable (Boltzmann), and smaller energy packets go hand in hand with smaller frequencies (Planck and Einstein), we should be desperate to know: "What determines the natural frequencies of systems?" Why do piccolos have higher notes than flutes?

The Orchestra of the Universe

The natural frequencies of repeated motions in the material world are well known (Table 3–1). The data show a striking trend: repeated motions become slower (ie are associated with lower frequencies) as the systems become more massive. As in an orchestra, bigger instruments have the lower frequencies – piccolo high, flute lower; violin high, progressively lowering from viola to cello to double bass.

Table 3-1
Natural Frequencies of Some Particles, and Aggregates of Particles[‡]

Particles or Aggregates	Frequency of Repeated Motion (times per second)
Electron in an atom	10^{15-16} (vibration)
Atoms connected in a molecule	10^{13} (vibration of the atom–atom 'bond')
Protein in a cell	10^7 (rotation)
Pollen grain floating on water	100
Wing of a bee	100 (vibration)
Wing of a sparrow	10 (vibration)
Wing of an eagle	1 (vibration)
Rotation of a human	1
Waving of a whale's tail	10^{-1}
Rotation of Moon round Earth	10^{-6}
Rotation of Earth round Sun	10^{-7}
Solar system rotating in the Milky Way galaxy	10^{-16}

[‡] *Increasingly large systems are listed upon progressing down columns 1 and 3, and their corresponding natural frequencies are listed by progressing down columns 2 and 4. The times taken for the repeated motions can be derived from the reciprocal of the frequency (eg a bee's wing takes about 1/100 of a second).*

The trend is everywhere. The natural frequency of the electron (only 1/2000th of the mass of a proton), as it moves within the atom is astonishingly large (ten thousand million million times per second). As the directly connected atoms of a molecule vibrate relative to each other, their natural frequency is about 1,000 times slower – simply because the vibrating atoms are more massive than the electron. The wings of a bee vibrate around 100 times per second, much more rapidly than those of the sparrow, and even more so in comparison to those of the relatively massive eagle.

A mathematical analysis of the natural frequencies of vibrating systems demonstrates that it is not only a larger mass that determines a lower natural frequency. The lower frequencies are also associated with weaker connections between the masses of the oscillating systems. The grand piano reveals all – the strings of its low notes made out of coils of wire, but its sopranos constructed from straight wire. The coiled wires, more easily extended than the straight wires, are aided in their descent to the depths by both larger mass and flexibility.

So, systems of larger masses, restrained by more flexible forces, are associated with lower frequency. Additionally, the prior section showed that smaller energy packets are those of lower frequency. Together, these conclusions establish an important rule:

> *Sloppily constrained systems of large mass go hand in hand with energy that is present in small packets.*

Important because it tells us what will be the direction of change in the Universe. And what could be more important than that?

The Direction of Change

The lesson from Boltzmann was that if there is a change between two possible states then the state in which the available energy is sub-divided into smaller packets is the more probable. *Ergo, the direction of change will be to pass energy from vibrating systems of small mass that are tightly constrained to systems of larger mass that are more loosely constrained.*

This realization is not esoteric – it has profound consequences for what happens in the Universe. A photon from the Sun carries a large energy packet, characterised by a high frequency (fast vibrations), to the Earth. So, when molecules on Earth absorb the photon, the energy in the photons is then spontaneously degraded into the energies associated with slower vibrations, and wider movements, of molecules (heat). That is why we warm when hit by photons of light from the Sun. The reverse process, that of gathering together all the small energy packets of our vibrating body molecules into one place to allow re-emission of the single high-energy photon, is extremely improbable. We cannot fire photons back to the Sun.

We have entered the core of all change. When plants grow, the change is accompanied by a degradation of some of the energy of the Sun's photons into heat. When animals eat to sustain their form, to grow and to function, the electrons that constrain the atoms within food molecules are reconfigured in new molecules (that are finally those within waste products) and some of their energy is degraded into heat. Heat is energy in small packets. Petrol combusts in a car, electricity heats water, a bomb explodes, food converts into faeces, tennis is played, a match burns – all the changes proceed in the direction that they do because energy is degraded into smaller packets; entropy increases. This universal law that covers all spontaneous changes is the Second Law of Thermodynamics:

In a spontaneous process, entropy increases.

It is a cornerstone in the enormous success of science. In 1959, C P Snow commented in his Rede Lecture ('The Two Cultures and the Scientific Revolution') that if non-scientists lack knowledge of the Second Law of Thermodynamics, it is equivalent to scientists not ever reading a work of Shakespeare; quite a condemnation.

Disorder Increases in Change

Many find the concept of entropy difficult. But even before Boltzmann led us to his statistical view, humanity was crawling towards an understanding of its importance in understanding the direction of change in the Universe. This earlier work derived from the efforts of a number of scientists who were thinking about heat and the degree of order in systems. Their approach is easier to comprehend.

In a shot at snooker or pool, kinetic energy (movement) that is initially possessed by the cue ball alone passes into several balls upon collision. These several balls eventually slow down and, finally, stop. Thus, in spontaneous change, energy that was initially in one place (the cue ball) first passed into several places (numerous balls). Eventually, it is "diluted" into so many places that it seems to have disappeared. In fact, the tiny molecules of the baize cloth and cushions eventually received the energy, with the consequence that there is a temporary, but very small, rise in temperature of parts of the table. The energy is still there, but we cannot spontaneously get it back into the cue ball – it has been "diluted" into a very large number of sites. This is Boltzmann's description – an increase in entropy.

But the snooker shot guides us to a second description of an increase in entropy. The energy that was initially localised in the cue ball eventually *disorders* billions of molecules in the baize cloth as they are warmed by collisions (heated). *Spontaneous change occurs with an increase in disorder.* Nobody has ever succeeded in putting all this dispersed energy back into the cue ball. This is a deep conclusion, because it tells us why we are unable to reverse time – time advances in the direction of an overall increase in disorder.[1]

So, if entropy seems difficult, take refuge in a simpler description. A bomb explodes, petrol combusts and a match burns, because the products of these changes are more *disordered* than the starting state. But if spontaneous change leads to more disorder, how did highly *ordered* planets, frogs and humans spontaneously evolve?

Why Cars Speed and Cockroaches Form

Clearly, increases in local order do occur; for a highly ordered crystal forms from a solution, and water is soon converted into ice if a glass of water is placed in a refrigerator at –5°C. Paradoxical as it may seem, when this local order springs spontaneously from a more chaotic state,[2,3] the Second Law is obeyed, not defied. For since the increases in order are *local*, it must also be considered what happens to the *surroundings* of that local environment. In the process of freezing, the water gives out heat to its surroundings and, in so doing, disorders them – and to a degree that is greater than the order created in the ice.[4,5]

The spontaneous ordering of water when it turns into ice is a change in the inanimate world. But the Second Law does not care whether matter is inanimate or animate; it treats rock and worms in the same way. In the general case, local order can be created providing that, at the same time, sufficient energy is degraded – passed into smaller packets – into the surroundings. This degraded energy commonly ends up as heat, and the associated disorder – due to increased agitation of the surroundings – more than compensates for any local order that accompanies the change. The localised energy provided to animals, poised for degradation, is food; that supplied to plants is almost universally photons.

Green plants harvest the energy of photons through photosynthesis. But photosynthesis involves very large and complex molecules. It could not have been the process used to organise and maintain the earliest forms of life. They may have possessed a more primitive way of harnessing the Sun's energy. Or they could have tapped the localised energy in molecules emanating from thermal vents or springs – sources still used in the depths of the ocean and near volcanoes.

In photosynthesis, the green pigment chlorophyll absorbs a photon – a large packet of energy. When the photon is absorbed, a negatively charged electron in the chlorophyll molecule increases its distance from the positively charged nuclei – energy has been stored in the modified chlorophyll. The electron, in a series of transformations, then gradually returns to positions that are successively nearer to the positive charge of nuclei. In this way, the original energy of the photon is released in a stepwise manner, as heat, to the surroundings. The consequent disordering of the surroundings more than compensates for the creation of local order – the regimented molecules and the flowers, stems and leaves that they form – in the plant.

So, as big packets of energy are degraded to smaller packets, *local* change can be the reverse of common experience. Through the absorption of photons, plants produce oxygen from water – rather than the normally explosive production of water from hydrogen and oxygen. Through the same absorption of photons, photosynthesis uses carbon dioxide and water to build plant sugars (with oxygen as a by-product) – rather than the burning of sugars in oxygen to give carbon dioxide and water. Through the combustion of petroleum, cars accelerate – rather than slow down. As big packets of energy are degraded to smaller packets, disorder increases *overall*, but *local* order can be simultaneously increased. The spontaneous generated order of trees, daffodils, cockroaches, monkeys and humans is not a problem. Photons and food – localised energy sources – allow change in a localised spot of the Universe to be turned on its head. The apparently miraculous is not so; rather a commonplace that is well understood.

Assembling Organisms

It's a very odd thing, as odd as can be, that whatever Miss T
eats turns into Miss T.

– Walter de la Mare

The witty observation of Walter de la Mare with regard to the effects of food – although, taking account of excretion, not strictly true – is striking. Who cannot be awed by the maintenance of a dog by repeated doses of only water and one kind of food pellet? At the surface of the Earth, the sequence of events in which energy is degraded as it passes from one life form to another is shown in Figure 3-1:

Figure 3-1. The passage of energy from the sun through the living world.

Since the first part of the sequence uses energy from the Sun then the existence of all the life forms that follow is ultimately reliant on this source. The plants derive their material form by using, with the help of photons, carbon dioxide and water as food. The wonders of photosynthesis can be demonstrated by a simple experiment. If a sapling of known weight is planted in a tub of soil, and the tree produced from its growth is many years later removed from the soil and weighed, an increase in weight of, say, 100 kilograms might be noted. Yet the change in weight of the (dry) soil in the tub is negligible compared to this increase. Almost the whole of the weight of the tree has come from the atmosphere. Rain and carbon dioxide have been converted to wood and leaves which, relatively speaking, contain only traces of nitrogen, phosphorous and minerals from the soil.

Grazing animals eat the molecules of which plants are made, and the energy in the plants is further degraded. In this way, the grazers organize and maintain their forms and are able to perform their characteristic functions. In the next step, carnivores eat the grazing animals, with a further increase in number of arrangements of the energy. In this last step, and in the competition for the energy supply, we can see that it is the laws of physics, chemistry and selection that determine a brutal aspect of life. Organisms, including humanity, dance to the tune of these laws.

The main by-product of photosynthesis is oxygen – for the plant world, it is the equivalent to the combination of waste carbon dioxide, urine and faeces found in the mammalian world. But the herbivores and carnivores then take up oxygen to "burn" their food. As a by-product, they exhale carbon dioxide. This waste product of animals is the "lifeblood" of the plants, just as the waste product of the plant is the "lifeblood" of the mammals. The whole series of events is a model of recycling efficiency.

Change and Equilibrium

When systems change, the overall amount of disorder in the Universe is increased. But when the overall degree of disorder has reached the available maximum, the system ceases to change. It is said to have attained equilibrium – as when physical change ceases when ice and water are in equilibrium at 0°C.

Equilibrium is also normally attained in chemical change – changes in the structures of atoms and molecules. In the Belousov-Zhabotinsky chemical reaction – named after the Soviet scientists who discovered and studied it – when the chemical reactants are mixed, there is initially an *oscillation* between two product states that differ in colour. The regular oscillation of the colour can continue for up to an hour, but it eventually dies down – all the reactant fuel has been used up and the system has reached equilibrium. Thus, oscillating states are properties of some chemical systems when kept away from equilibrium, and are not properties confined to the living. Periodic changes in the concentrations of chemicals are used to bring about periodic changes in organisms – the desire to eat or sleep; the time of menstruation; the beating of the heart. The beating of that traditional soul, the heart, lies in a chemical principle evidenced in the inanimate, but exploited in the living. If a system, living at one instant, attains equilibrium at the next, it is then dead.

When a chemical reaction occurs, the starting material is in one energy valley, and the product of the reaction in another. There is a mountain pass (energy barrier) to be surmounted to get from one to the other. The key question is: "Is the top of the pass accessible so that the reaction can occur?" A wooden table is observed in its most stable available equilibrium state, and the pass that must be surmounted to make it change is very high. However, if the temperature of the table is greatly raised, the pass is attained through a lot of energetic shunting of its molecules; products are then formed with descent into a deep valley. The table becomes a raging inferno with generation of new combinations of the table's constituent elements – those stable bits and pieces of carbon dioxide, ash and water.

Through various mechanisms (most obviously an increase in temperature), changes in local conditions allow access over passes that could not formerly be reached, or are attainable to different degrees. In both the inanimate and animate worlds when conditions change, the concentrations of molecules can therefore change. Given these programmed changes in concentrations of molecules within a daffodil bulb as a function of time, it is possible to trigger new events. Science taught us how daffodils can bloom in spring.

Subtlety is exercised (Figure 3-2) not only through the height of the energy barrier to be overcome to convert one system (circle) to another (oblong), but also through the extent of the following descent – greater in this case to produce an alternative system (square). In starting at the left, molecules will obviously tumble more frequently over the lower barrier to make oblongs rather than squares. But although few squares are initially made, they less frequently revert to circles – due to the higher mountain pass they must traverse – than do the oblongs. So in the long run, given all the competitive backwards and forwards traverses, the system of squares becomes more highly populated than that of oblongs.

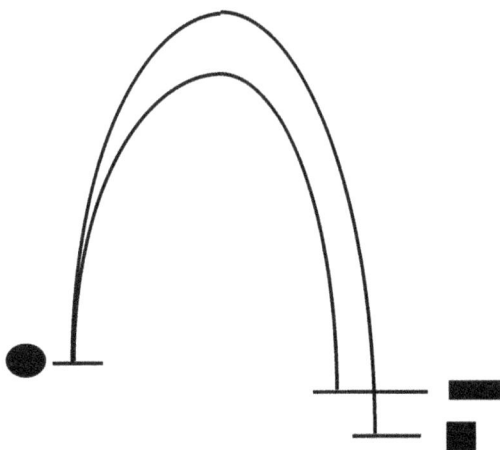

Figure 3-2. Subtle variations of time or temperature can produce new and different systems from a common starting point.

So, oblongs are initially the main products, but squares later dominate. By exploitation of these principles, Nature has found us to be possible, and holds us in awe. Menstruation can be held in limbo for a decade; then occur monthly for a few decades before dying away. And, since an increase in temperature causes molecules to be more energised, given time they can then pass more quickly from a system represented by a circle to one of squares in preference to oblongs (Figure 3-2). The otherwise astonishing appearance of females, or males, from the eggs of some reptiles,[6] and a few birds, according to the temperature at which the eggs are incubated, becomes comprehensible. Chemistry exercises its principles whether working on the inanimate or in dictating the requirements of life.

Requirements for Life

Bread is the staff of life.

– Jonathan Swift (1667-1745)

All living systems have striking features in common. First, they are kept away from equilibrium with their surroundings through the frequent intake, and turnover, of energy. In some cases, organisms can postpone the intake and turnover of energy though the process of hibernation; or be forced to postpone it through cryogenic techniques. But all systems that are actively living are characterised by continuing changes that are energy-driven.

Second, living systems are copied by passing on complex information coded in molecules. Third, this information, in conjunction with information and other influences from the environment, controls both the organization and functions of the living system. Which is why an understanding of molecules, and of the spontaneous changes occurring in them, is crucial to an understanding of what we are, and what we can be. So, the key molecules selected in the journey of matter that give rise to buttercups, bugs, butterflies and bears – and of course, humans – are considered next.

4

The Factories of Life

I see an extraordinary potential for human betterment ahead of us.
We can have at our disposal the ultimate tool for understanding
ourselves at the molecular level…The time to act is now.

– James D Watson, April 1988

The oldest definitive macrofossils are dated back to about 3,200 million years ago. And their prototypes probably evolved within a billion years of Earth's formation. So the factories of life have had a long time to become sophisticated.

If we are to understand ourselves, we must understand something about these factories and how they work: the bits and pieces from which they are made; how they get their walls; how they pass on information to make new ones; how they make their products; and how these products do their jobs. And, where possible, how these factories came into being. In sum, we must understand at least a little of how the molecules, from which we are forged and through which we function, work.

The Common Bits

The organic molecules that are the basic constituents of life vary greatly in size, but many are about 10-1,000 times as long as an atom. The size variation naturally elicits the question: "How are larger molecules made?" They are made by pathways – beyond our present scope – that are understood in great detail;[1,2] and always in accord with the principles covered in the preceding chapters. Where complex, these pathways co-evolve with the complexity of the organisms within which they operate.

Rocks that are about 3.5-4 billion years old – around when organisms first appeared – are also found to contain rounded pebbles; an indication that they were constituted from water-ground rock. So, water was probably around at this time, indicating that the temperature of at least parts of the Earth's surface was not very different than from what it is today. Water appears to have always been a key molecule of organisms. In those of today, it is the most abundant; fruit and vegetables with up to 95%, humans with about 65%.

But evolution has called upon many of the elements; humans contain about 60 of the 100 bits of the available Lego kit. Yet the vast majority of these are used in small amounts – so much so that 96% of our content is made of only four: C, H, N and O, perched at the top of the Periodic Table. Nature is impressively parsimonious in the parts of its Lego kit that it mainly uses. When these "little guys" combined together, they could produce molecules that stuck together with just the right amount of cohesion to allow the development of flexibility, mobility and complexity – all key features of life.

During a few billion years, such molecules were fashioned into four major types: the carbohydrates, here ignored at not too great a cost (a diet that permits a book of limited obesity); the fats, which notably make the containers for life (membranes); the nucleic acids (DNA and RNA) which code for the construction of life's bits and pieces; and proteins, which are the units for construction and function. These common bits are universally packed in cells.

The pathways for the construction of DNA, RNA and proteins are complex. But biological complexity cannot arise instantaneously – it must be selected. So the earliest molecules of life must have been enormously simpler. The problem of gaining knowledge of these much simpler precursors has proved insurmountable. And we should not be surprised, for even the fossil record of organisms constructed through the agency of DNA, RNA and proteins (that is, *all* known organisms) is sparser than we would wish; even their hard bones and shells succumb except in environments that are protected.

And we cannot guess what the simpler precursors were. Analogies illustrate the problem: someone exposed to communication *via* a word processor and a laser printer, but isolated from the historical record, would find it difficult to deduce that quill pens had been used in earlier times. Without historical record, the form of the steam engines that preceded diesel engines is not self-evident. This "sweeping away" results in a large discontinuity in the transitional states between inanimate and living systems.

Those with the God-of-the-Gaps syndrome, loving "missing links", may concentrate their attentions on this discontinuity. But they should be aware that there is no discontinuity of principles. The existence of living organisms is perfectly consistent with all our scientific knowledge – as illustrated in chapter 3, no need for magic. The systems known as "living" could gradually evolve from the "non-living". With the passage of time, a limited set of molecules (from an essentially limitless pool) was generated and selected to store information, pass it on, and through this information build the forms, and perform the functions, of organisms. So when it comes to molecules and their aggregates, we must deal largely – but not exclusively – with today. And, since the molecules of life are all parcelled into cells, we start with them.

Cells

Cells are not only the hectic factories of life but also its smallest functional units – ranging in shape from spheres to lengthy strings. The tiniest are about the same length as 100,000 atoms – about 1/100th of the width of a human hair; the largest – an unfertilised ostrich egg – has a diameter of about 10cm. They are impressively small compared to the factories made by humans; and enormously more sophisticated – to levels that astound.

Bacteria consist of a single cell type, although these single cells often aggregate. In a human, there are about 200 clearly different cell types (liver, kidney, etc). But such is the multiplicity within these types that it is estimated that, in total, there are somewhere in the region of 100-10,000 trillion cells in total – and similar numbers in our nearest mammalian relatives. The estimate for total number of cells is woolly, in part because it is so gargantuan – at least 10,000 times larger than the global population of humans.

In terms of the total number of cells forming our systems, we are only about 10% inherited from egg and sperm. The remaining 90% are our busy partners – bacteria. So numerous that they constitute 30% by weight in dry faeces; that around 10,000-1,000,000 of them make their home on a typical square centimetre of skin. Moist areas – the groin, nostrils or between the toes – harbour much larger numbers. We cannot survive without them. Vitamin B12 is necessary for the formation of blood, and for normal function of the brain and nervous system; our cells cannot manufacture it. In our guts, the factories of bacteria produce it effortlessly; they are sophistication personified, for the laboratory synthesis of vitamin B12 took years of human chemical toil.

That cells are the key universal element of all living systems is not surprising, for, to pass on information and to provide function, molecules must recognise and therefore find each other. To meet this requirement, it was useful to have all the miniscule goodies restrained within a tiny bag – the cell membrane.

Cell Membranes

The cell membrane can be likened to much larger objects: the walls of a room, a shopping bag or the surface of a soap bubble.

Nature makes these cellular bubbles through the aggregation of very large numbers of snake-like molecules. Each of these molecules prefers the configuration of a rod, but is flexible. The molecules are like fat along their length, but at one end have a "head" which behaves like table salt. Being salt-like, this head quite happily mixes with water; but, as with paraffin, the fatty part prefers not to mix with water. This dilemma is resolved through spontaneous aggregation of very large numbers of the snake-like molecules to give a structure like a soccer ball. Fatty portions are hidden between the inside and outside surfaces of the sphere (Figure 4-1). The salt-like heads are distributed at the inside and outside surfaces – which both interface with water. The membranes of the cells of *all* living organisms – although containing additional components – are constituted in this way.

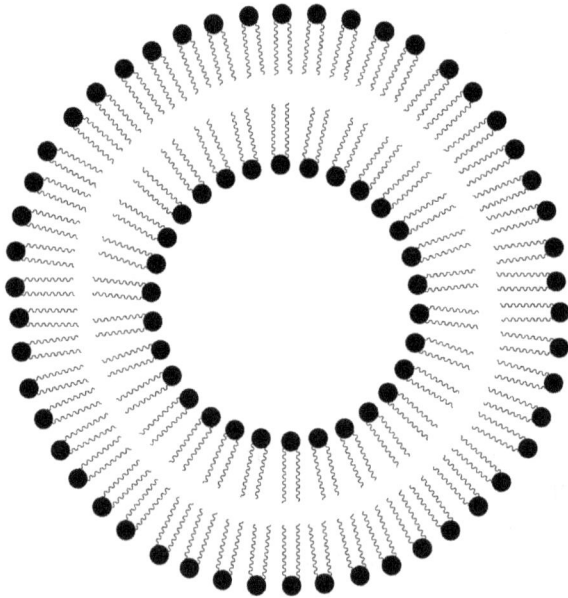

Figure 4-1. A cell membrane – the universal container for life: the salt-like heads (circles) of individual molecules form the inside and outside boundaries of the membrane, and their fatty portions (black) are sequestered in its interior. The basic units of life are constrained inside the membrane, but both inside and outside the membrane the most abundant component is typically water. In this simplified model, the thickness of the membrane relative to the diameter of the cell is grossly exaggerated.

The cells (prokaryotic) of early fossil micro-organisms, and present-day bacteria, possess only this boundary membrane. The cells (eukaryotic) of today's plants and animals possess, in addition, internal membranes. One of these internal soap bubbles – the nuclear membrane – restrains almost all their genetic material (DNA). Bubbles, like cell membranes, do coalesce, but getting one inside the other is not easy. So how did membranes get inside membranes?

Membranes are more subtly plastic than bubbles. Around 1.5 billion years ago, one cell swallowed another, and a membrane inside a membrane evolved. Eukaryotic cells came into being; the ancestor of plants and animals had been born. Lynn Margulis first proposed this imaginative scenario, now amply supported by much evidence. In the same way, artificial membranes (liposomes), readily put together in the laboratory, spontaneously engulf one another. There is no discontinuity between the physics and chemistry of inanimate and living systems. And physics and chemistry demand that, to be able to provide function, the contents within the cell membrane must harvest energy.

Cells and Energy

A power station takes in coal or oil – fuels then converted to the common currency of electricity that drives the organisation and functions of human societies. In the same way, cells maintain their organisation and functions by harvesting and then degrading energy (chapter 3). One way or another, they must acquire this energy from their environment: store some for a rainy day; convert some to a common currency for early use.

There was no oxygen in the early atmosphere. So the early bacteria-like organisms were anaerobes – they derived power without using oxygen, perhaps from molecules produced in hot springs and vents. But, with the passage of time, some cells (prokaryotic cyanobacteria) evolved the ability to trap the Sun's photons after their journey of 96 million miles. These solar-powered cells were later swallowed by a cell that had hitherto had taken its power from the local environment. The route to green plants had opened up, for the photosynthesis of today – with its associated production of oxygen – evolved from the invaginated cyanobacteria.

Once the green plants got going in abundance, the anaerobes had a problem – to them oxygen was toxic and threatened their demise. The laws of Nature had led to "pollution" on a global scale – in this sense, nothing new in the CO_2 that is the possible culprit in global warming. But the descendents of the anaerobes are not only still around – they are necessary for our survival. What were their survival strategies?

One strategy was to hide from this dreadful, life-threatening oxygen. They withdrew to niches where oxygen was absent, but where their favourite dishes were still available: to deep marine environments near to gases escaping from the Earth's interior; to within marshes. Today, their ancestors occupy rubbish dumps and the human gut; and are responsible for repulsive smells. With a burgeoning of the population of humans, and their associated garbage, real estate containing their dream homes is increasingly created.

A second strategy was to detoxify this dreadful, life-threatening oxygen through the evolution of enzymes that removed its toxic by-products. And, as the plants further polluted the environment of the anaerobes, these evolving opportunists actually got to rather enjoy oxygen. They exploited what had been a pollutant, evolving pathways to use it as a part of their source of energy. The anaerobes had become aerobes.

A third strategy was now possible: any anaerobe having problems with oxygen could simply swallow an aerobe. As with the regurgitation of Jonah by the big fish, the engulfed soul was, on occasions, spewed out. But the cargo was too valuable to be always thrown overboard. The organism of smells could use the engulfed aerobe to provide energy by burning food molecules through reaction with oxygen. Those aerobes are the ancestors of the mitochondria of today – the miniscule power stations of animals in most cases passed on to offspring through the mothers (as for ourselves). For given a few billion years, the combination became gloriously advanced, differentiated and multiplied into collections of trillions. It was the trick that led to humans and other animals. We have a smelly distant ancestor made wholesome by our oxidising factories for energy production. These mitochondria exist, with their own DNA, inside all our cells – except for the red blood cells. They use the energy from food to make a molecule that is almost the universal energy currency of organisms (ATP). Raise a hand and you use countless billions of molecules of ATP. Thank heavens for little bacteria.

The origins of our mitochondria show how advance is often well served by friendly cooperation. But often it is also served by brutal competition, for there is an alternative strategy to harness energy. When one form of life encounters another, it simply eats it. Just about every living thing, except the benign plants, gets up to this trick. The simplest way that a single cell organism can "eat" is through invagination. In the first stage, a small part of the outside surface of the cell membrane, containing a food particle or even another cell, is folded into the internal part of the cell – rather like pressing your finger into a balloon. This small invagination then buds off. A basic requirement of life – energy provision to the interior of the cell – has been met. But the plasticity of membranes provides a mechanism not only for nourishment; it also provides the means for the transmission of life itself.

Reproduction

The plasticity of cell membranes allows them to be squeezed inwards, the whole way round the middle of the cellular ball – as in squeezing a balloon to produce a dumb-bell. Imagine a complete cut at the middle of the dumb-bell and, if the two ends are synchronously sealed, two cells are produced from one. Squeezing, splitting and sealing is facile for membranes – the method by which asexual reproduction occurs in bacteria, plants and fungi. Sexual reproduction uses the same dodge – nip, cut and seal – when an egg cell divides following fertilization. The daughter cells, produced either from vegetative reproduction or from egg divisions, carry a copy of the genetic material of the parent cell, and thus information accumulated over previous generations is carried forward.

Nip, cut and seal is an energy driven process. As such, energy can also be used to press it to go in the other direction – abut, cut and fuse. Sperm abut eggs; enzymes next cut intervening membranes; DNA of sperm and egg are then fused. The ductility of membranes is an enabling feature of reproduction. Abut, cut and fuse (fertilisation), followed by countless nips, cuts and seals (development), occur when sperm and egg lead to a baby. These processes do not create life but are part of the processes through which it is copied and modified. But how is the copying achieved? Jim Watson and Francis Crick provided the details.

Life's Instructions

Francis Crick: *When the war finally came to an end, I was at a loss as to what to do...*
Later: *We've discovered the secret of life.*

Francis, along with his colleague James Watson, found something important to do, and their work resulted in something supremely important – the structure of DNA. DNA molecules are long strings; information is carried along their length. With the exception of red blood cells, they are found in all of our cells; indeed, in all organisms. The backbone of the string provides a long scaffold to which the information-carrying molecular motifs are attached. The motifs are four "bases" with shorthand names **A, T, G and C**. Each of the bases is constituted from fewer than 20 atoms of the elements C, H, N and O.

By simply heating a few simple chemicals together in solution, the base **A** can be synthesised with surprising efficiency. So, in principle, the bases could predate the organisms that now produce (biosynthesise) them. However, DNA strings are complex, and clearly a product of biology. The more primitive systems that allowed information storage and limited function have been swept away. We can only guess at what they were.[3]

The DNA strings of today are models of beauty and economy. If a ladder were cut vertically through the middle of its steps, then the resulting long vertical halves would represent the DNA backbone; if encoded with red, blue, yellow and green paint, the "half steps" would represent the differing bases. The order of the four bases along the ladder encodes the information for the construction of living organisms.

How can the information be copied? When around 1440, Johannes Gutenberg used shapes (letters of the alphabet) to mould their complementary shapes, his pioneering printing press forged the complement each time it was used. Nature works slightly differently. By about a billion years ago, and possibly even 3-4 billion years ago, it had moulded, and could continuously make, two complementary pairs. The complementary fits are **A** to **T** and **C** to **G**. Once the complementary pairs have formed, they are represented as **A...T** and **C...G**.

The fits are based upon complementary shapes – as in Gutenberg's printing press – and local attractions of opposite electrical charges. But it should be added that, as for our sense of touch and impact, the only ultimate determinants of the **A...T** and **C...G** recognitions are electric charges; for when a medicine ball impacts mightily on your frame, the perception of its shape occurs through its electrons refusing to invade the boundary set by yours. Likewise, when you hit a brick wall at speed, bones and flesh are displaced due to the refusal of one set of electrons to invade the space of others. Electrical forces determine the things we call "shapes".

So, in Nature's copying, the **T, A, C** and **G** bases on an existing string respectively recognise separated **A, T, G** and **C** bases individually available from solution. Hence, given that these bits from solution are stitched together as they are recognised by the pre-existing string, a complementary string is made. An original sequence **T-A-C-G** would give a complementary string **A-T-G-C** (called copy-DNA, or c-DNA). Initially, the c-DNA sticks to the template from which it has been forged, with the two strings – one old, one new – wound one round the other; a twisted ladder – the famous double helix of television fame.

In announcing the discovery of the double helix, James Watson and Francis Crick implied, with masterly understatement, that the biological implications of their structure had not escaped them. For dissociation of the double helix into two single strings frees the c-DNA sequence, which can in turn be used as a "reverse template" so that a new copy of **T-A-C-G** is produced. This is the mechanism whereby – given the copying of enormously long strings of DNA – information is passed on to new cells within a body; to make livers, kidneys, lungs, legs and toenails; robins with the red breasts of their parents; children carrying features of their parents. The Gutenberg Bible was a copy of the words of Man, believed accorded by God. A new string of DNA is a copy of the words accorded to organisms.

The discoveries of the molecular mechanism of heredity, and of an expanding Universe, arguably rank as the most momentous discoveries of the twentieth century. For their discovery, Watson and Crick were awarded the Nobel Prize in Physiology or Medicine in 1962 (jointly with Maurice Wilkins who, with Rosalind Franklin, had provided crucial experimental evidence that supported the proposed structure).[4,5] Sadly, Rosalind Franklin lost her life to cancer before her important contributions were widely recognised. The double helix is indeed one of the greatest secrets of life.

But does it not amaze that sequences of just four bases can signal the likeness of daughter to mother? Proteins – biology's components – are the direct products from DNA that enable such seeming miracles. How are they made – how is memory converted into function?

Life's Components

A cell full of just DNA could not get up to useful tricks – all memory, no action. In the cell, as in business: there must be a product. And, the better it is, the more likely you are to be successful. Henry Ford knew this and, in his production lines, knowledge accumulated over countless generations was translated into machines that could do things – automobiles. Biology's accumulated knowledge – DNA – is similarly translated into function on a production line. The products – proteins – have a fascinating similarity to automobiles for, as this name implies, they are automated in their functions and mobile in exercising them. Once more, the molecules were ahead in the game. Henry Ford followed a few billion years later.

In the very complex systems of today, the translation of memory

(DNA) into functions (proteins) does not occur in a single step; it goes DNA –> RNA –> protein. Although this complication is sophisticated in its details, it can largely be ignored here. For RNA is similar to DNA, typically also having four bases strung out upon its length. How is the transformation RNA –> protein achieved? The production line, the apparatus carrying out this function (the ribosome), is similar in all organisms. As for the automobile productions lines of today, it is inevitably complex. Its evolution to its present state must have taken a period at which we can only guess – perhaps a few hundred million years?

A couple of twentieth-century devices, cine-projectors and tape-reading computers, read long tapes, fed in at one side and exiting at the other. From the information on the tapes emerged useful functions: either a motion picture, or tasks carried out by the computer. The ribosome works in the same way; but once more, biology pre-dated humanity's devices by a few billion years and worked at less than one millionth of their scales. The RNA tape is read in at one point on the ribosome, and exits at another. The information on the RNA tape (in the sequence of four bases) instructs for the making of proteins – biology's Lilliputian machines that are assembled from linear strings of 20 building blocks – the amino acids. It is not clear why Nature homed in on 20 building blocks – rather than say 18 or 22. But, given that it did, how can strings containing 20 different things (amino acids) be made from strings of just 4 different things (bases)?

The Genetic Code

The elucidation of the genetic code is indeed a great achievement…
It is, in a sense, the key to molecular biology…

– Francis Crick

The genetic code is a wonder of the logic that emanates from selection. Clearly, 4 different bases, taken individually, can only specify 4 different amino acids. But, since science does not bow to miracles, it turns instead to the principles of mathematics. In order to generate the larger number of amino acids, the bases must be taken as *combinations*. A child might solve the problem. Given only an apple (**a**), a pear (**p**), an orange (**o**), and a lemon (**l**), but asked to produce 20 linear patterns (where pear-lemon is different to lemon–pear), it might first try pairs of the fruits. But this gives only 16 patterns: **ap, pa, ao, oa, al, la, po, op, pl, lp, ol, lo, aa, pp, oo, ll**. However, if the fruits are taken in groups of three, there are 64 combinations. Thus, if the bases are read from the RNA tape in groups of three, there is more than enough variety to code for 20 amino acids.

And triplets were exactly what Nature chose. A sequence of three bases specifies the amino acid that must be incorporated at any point in a protein. Gratifyingly, at the levels of astronomy ($E = mc^2$), biology and children, the logic of mathematics is applicable. And RNA sequence must tell where on the tape protein synthesis must stop – tell the production line when it should grind to a temporary halt. So, three of the triplets do not code for an amino acid, but instruct: "Stop the synthesis of this protein – the required biological device has been made." The production line also needs to know where the synthesis of a protein must start. Two triplets are used for this function. These two also code for amino acids, which may later be removed from the beginning of the protein molecular string if unnecessary for its function.

Since 61 of the 64 possible triplets code for the 20 different amino acids, several different triplets specify the same amino acid; there is degeneracy in the code. Lest degeneracy be thought inefficient, contemplate one possible alternative solution. A doublet code could code for a maximum of 16 amino acids. If, in addition, four amino acids, and START and STOP, were coded for by a triplet code then coding for a total of 20 amino acids is possible. But two production lines would be needed. Henry "any colour you like as long as it's black" Ford would surely have baulked.

Failure to make the translations stipulated by the genetic code could be fatal. Yet, until old age, we rarely live on a knife-edge. Errors – themselves relatively uncommon – are corrected. Everywhere, we see the code at work. For, among the species (around 2 million are currently classified, and estimates of the total number range from around 5-50 million),[6] its use is essentially universal, not only in the making of humans but also in the making of bacteria, crocuses, mushrooms, oak trees, venomous spiders, parasitic wasps. It is another grand unification.

Francis Crick, Sydney Brenner and their colleagues in Cambridge indicated that the genetic code was read in triplets. Robert Holley, Gobind Khorana and Marshall Nirenberg were awarded the Nobel Prize in Physiology or Medicine in 1968 for their work that established its details. But what are the forms of the bodily components that are stipulated by DNA?

Structure and Function of the Components

The 20 amino acids used to build biology's components are themselves made from relatively few atoms of C, H, N, O and sulphur (S). They are rather floppy – capable of taking up a variety of shapes. Some arrived on Earth from meteors; many can be made by stewing up a few simple chemicals together in the presence of energy sources (eg electrical discharges) that would be available during the early history of the Earth. But in the biological systems of today, organisms make the amino acids. The 20 are here represented as a,b,c...t. In the proteins, stipulated by DNA –> RNA and the genetic code, the 20 have become attached together in long strings. In the strings, they can – and commonly do – occur more than once. The minimum lengths are about 70 amino acids (say atrsceghhbnmmks............qcdd). Lengths of up to several hundred are common; the longest known (in striated muscle) has about 34,000 amino acids.

The number of possible long necklaces that could be constructed from beads of 20 different colours, where the colours can be used more than once, and where the order of the beads is important, is gargantuan. But Nature obviously does not acquire sophistication through the production of alphabet soup, any more than engineers could construct useful devices from a random assembly of objects. Nature's useful devices require very special amino acid sequences, just as a jet airliner requires thousands of components to be assembled in the places where they can give appropriate function. Just as society needs hundreds of thousands of objects to provide a sophisticated whole, the structural variety of the proteins that endow organisms with forms and functions is enormous. Current estimates are that translation of the human genome leads to around 100,000 to 1,000,000 proteins. But these are no more than educated guesses – nobody knows as yet. Nothing like this number of miniscule devices has yet been identified.

Nature selected 20 amino acids as the building kit from among millions of possibilities. In selecting these 20, and then joining them in highly selected long strings (proteins), Nature produced necklaces that could then spontaneously fold – as would real necklaces when placed in a tiny jewellery box. The nature of the folding depends on the nature, and sequence, of the amino acids in the necklace. Structures that fold were selected because folded structures are able to produce construction materials and mini-machines. Some are like flexible rods; others like miniature balls of wool.

Why may proteins spontaneously fold? The long molecular strings relinquish their extended structures by satisfying the weak attractions within themselves – those attractions that similarly produce liquid water and ice from water vapour. The small net charges ($\delta+$ and $\delta-$, chapter 2) within the strings satisfy the demands of the electromagnetic force when they become proximate in the weakly bonded interaction $(\delta+)$--$(\delta-)$. Hence, the string. It folds to give the more the more stable hairpin:

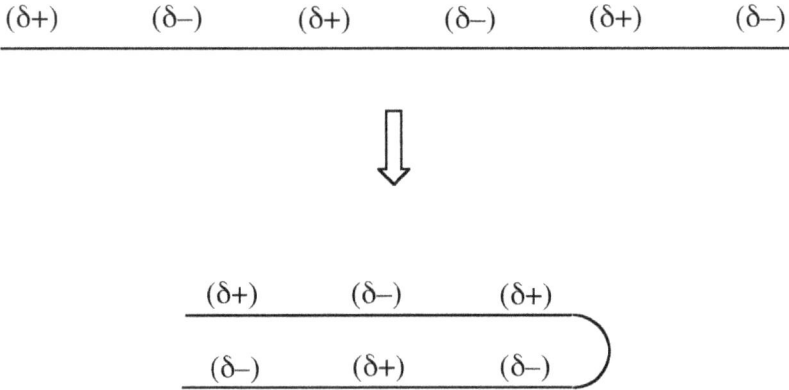

Figure 4-2. Charge-induced folding of a string.

These molecules, with their folds of hairpins, sheets and coils, may loosely be likened to the body parts of a car, and to its controlling structures of clutch, brake and accelerator pedals. But the proteins are enormously more sophisticated than are the parts of cars. Their forms are the solutions – automatically found – to three-dimensional jigsaw puzzles. Sometimes they collect together in swarms to provide construction materials for the body. Sometimes they acquire smaller molecules – oxygen, a food molecule, a hormone or a vitamin – that nudge them into new shapes. They are the stuff of skin, hair, nails, connective tissue and muscles. The astonishing variety of their textures – from squishy to hard – depends on the nature of their folds and the way in which they may aggregate.

Their apparent magic shows when they act as enzymes. Without these catalysts, there is – literally – an insurmountable barrier to the conversion of food into the energy required for bodily functions. As seen earlier, such barriers to change are almost universal: to the change of one element into another; of petrol into gaseous products; of a daffodil bulb into a flowering plant. Without them, everything would quickly descend into the deepest hole (energy well), and an interesting Universe could not have evolved.

To organisms, the barrier to the metabolism of food to release energy was a real Everest. Table sugar appears inert, and shows no sign that, if eaten, it will provide energy to the organism through its transformation into carbon dioxide and water. A similar problem is faced by a freezing human with a supply of wood but no source of heat to initiate a fire. The combination of oxygen from the air and a wooden table cannot be used as a source of energy until thermal energy is provided to get over the barrier to convert it to the new system of carbon dioxide, water and ash.

To solve the problem of food metabolism, biology could not resort to the crude method of raising the temperature to provide the thermal energy to get over the barrier. The high temperature would be incompatible with maintaining other delicate interactions that are necessary for life. What other solution might be possible? Well, given an athletic competition for children where, for the hurdles event, adult hurdles had been provided, the organisers might cut off a portion of the legs. The protein enzymes that catalyse the use of food arrived at the same solution – they lowered the barrier.

So, with the coming of life, there was the gradual evolution of a multitude of molecular devices. The first detailed knowledge of the structures of a couple of these devices, with their characteristic motifs of folding, was due to John Kendrew and Max Perutz. Their images of the structures (of myoglobin and haemoglobin) were obtained by X-ray crystallography, in an endeavour that required years of patient study. Their images exposed the coils and bends of nature's devices, built on a scale more than a thousand times smaller than can be seen in a microscope. Many had regarded their goal as impossibly difficult. They were finally rewarded when they shared the Nobel Prize in Chemistry in 1962.

Proteins are the devices, honed through blind forces, which enable our functions. They are the major stuff of humanity, although this stuff must still travel a long way to give the mobs of enormous complexity found in a brain. But we already know that the proteins are so numerous that a lot of DNA is required to code for them. How is all the necessary information to make all these devices packaged?

Genes and Chromosomes

The material at the core of a plant or animal cell – the nucleus – absorbs coloured dyes; and because of this property was named, from the Greek, chromatin. When one cell divides into two, the chromatin disappears and is replaced by rod-like structures – chromosomes. It seemed chromatin and chromosomes were closely related and, since this change occurred when cells were dividing, that they might be involved in passing on the heredity information.

The simplest idea is that a specified length of DNA (a gene) codes for a specified protein. But complex organisms are coded for by tens of thousands of genes, so all the genes are – in the interests of getting as much as possible in the box room – tightly packed within a much smaller number of chromosomes. When two copies of the chromosomes are available within a cell, it can divide into two cells and each of these product cells can contain one copy of all the hereditary information. The spontaneous generation theory of life – which had so dogged advance – had finally bitten the dust. It was Thomas Hunt Morgan who, through investigations of inherited characteristics of fruit flies, proposed and established the chromosomal theory of heredity, and that genes are situated on chromosomes. For his discoveries, he was awarded the Nobel Prize in Physiology or Medicine in 1933.

Fruit flies are small and have a relatively short generation time – conveniences ensuring their role in many more key genetic discoveries. The benefit of "blue skies" research is indicated. It is a misconception that genetic experiments with worms, flies and zebra fish constitute "ivory tower" research. The principles, and practices, of life are so universal that fundamentally important discoveries are made through their study.

Diversity through Combination

In bacteria, a contiguous length of DNA is indeed translated to give one protein; "one gene gives one protein" holds very well. However, in more complex organisms (eg mammals), a corresponding length of DNA is, for the process of its translation, divided into segments (eg A, B, C and D). Some of these segments are discarded, and the remaining parts recombined with segments from other genes. In this way, a greater variety of proteins is generated from relatively few genes.

The worm C. elegans (whose genome was the first of an animal to be almost fully sequenced) has around 19,000 genes, with the implication that it produces at least this number of proteins. The fruit fly, much loved by geneticists, has around 13,000 genes. The number of human genes is subject to some uncertainty, but there is evidence to support around 20,000. Yet a human – or indeed a fly – is clearly more complex than a worm. This greater complexity appears to be achieved by joining segments of genes together in a combinatorial manner, so leading to more protein devices.

It is one thing to describe the structures of the main molecules of life, and to understand a few details of how complex molecules are made in an automated way; another to understand how devices are made from them. Once more, guidance comes from the automobile.

From Molecules to Devices

Give us the tools, and we'll finish the job.

– Winston Churchill
to Franklin D Roosevelt

When a car moves, there is the consumption of fuel, and the familiar transmission of signals between one part and another. The transmission of signals is possible because one component fits into another. Organisms work in the same way: they consume food and also transmit signals between the component parts. Their millions of different molecules are able to signal to each other because they "recognise" and "fit" to each other – in a manner that is analogous to the way in which an ignition key "recognises" the ignition lock.

It was the great German chemist Emil Fischer (born 9 October 1852, near Cologne) who, in 1890, proposed the recognition of one molecule by another in his now famous "lock-and-key" hypothesis. He carried out monumental work on the structures of sugars, and made fundamental contributions towards early knowledge of the structures of proteins. For his work, he received the Nobel Prize in Chemistry in 1902. Emil and his wife Agnes had three sons: one killed in the First World War; a second took his life while undergoing compulsory military training. These tragedies probably contributed to the reasons for Emil Fischer's own suicide in 1915.

Fischer's "molecular recognition" is the essence of the mechanism by which organisms function through the use of proteins, although, in one way, it is more sophisticated than his original concept. For, through the recognition of one protein (more generally, molecule) by another, a change in the shape one (or both) may occur. This principle allows a molecular signal received at one point to produce change (eg movement) at another point in biological systems. Such changes also occur in molecular interactions divorced from biology. Movement within molecular structures follows the same principles in both animate and inanimate systems.

So, the molecular world of biology does not simply have keys that fit locks; it also has keys that change the shapes of the doors containing the lock. It is changes in the shapes of proteins (muscles), induced by molecular keys, which allow animals to move. There are proteins that can be induced in this way to snap shut like a gin trap, open like a jackknife, or assemble into long filaments.[7] The movements of muscles are no mystery. Nor do "mechanical forces" demand an addition to the fundamental four forces of chapter 1, for both muscles and pistons use forces generated through the electrostatic force as metabolism and petrol combustion occur.

The principles governing changes within the devices of proteins are those covered in chapter 3. These same principles operate when "magic gel" hand warmers are bent to trigger crystallisation of their contents, and hence to induce their warming cycle.[8] The contents of the pack can be returned to their gel state (by putting them in boiling water for a few minutes) and hence used repeatedly as a source of warmth. The use of hot water is equivalent to the use of food by an organism. Gel or organism, there is a controlled and recyclable source of change and energy. The existence of life is not mystifying.

But it is very impressive. The smallest known genome consists of about half a million bases; the human genome of just over 3 billion. But only a few per cent of the latter is translated into proteins. This non-protein-coding DNA was once dismissively referred to as "junk", and regarded as "hitching a ride" – in the same way as a parasitic organism uses its host species. But, from another viewpoint, its presence seems to make no more sense than coding the hard drive of a computer with memory that was never to be used. The traditional "junk DNA" is now under close scrutiny. By 2007, function had been discovered for some of the DNA formerly regarded as junk.[9] Once more, the story starts with blue-skies research.

In the 1940s, the American geneticist Barbara McClintock was researching the genetics of plants – particularly maize. Hardly a topic that, if research is being directed from the top of the pyramid of power, would excite a politician or a government official in control of the purse strings. But Barbara thought in a remarkably clear manner and had a habit of getting to the heart of the matter. In her experiments, she noted that, on occasions, control elements (stretches of DNA that control the expression of genes) sometimes changed their locations in the plant genome. In other words, she showed that DNA could jump round in genomes, even from one chromosome to another.

These stretches of jumping DNA became known as transposable elements. Nearly half of the human genome, formerly simply regarded as "junk", is composed of such jumping DNA; for principles that operate in maize also operate in humans. Even though these stretches of jumping DNA are not translated into proteins, they influence the evolution of the numbers of copies of genes in the genomes of organisms. The expression of genes is affected by transposable elements, and therefore they influence the nature of organisms.

Barbara McClintock's work was important to her: "I know [my corn plants] intimately, and I find it a great pleasure to know them." It was also important to the world of organisms, and Barbara was awarded the 1983 Nobel Prize in Physiology or Medicine. By 2010, it was shown that much DNA traditionally regarded as junk, although not translated through to proteins, is transcribed to RNA, and that this RNA often has a function. As Sydney Brenner commented, "There is lots of useful stuff in the attic." But what is our inheritance outside the standard will of the genome?

Heritable Changes Outside the Code

Before inheritance was understood in terms of DNA sequences, Jean-Baptiste Lamarck (1744-1829) proposed that "acquired characteristics" could be inherited. The big muscles of the body builder are an "acquired characteristic", but do not occur through *changes in base sequences* within genes. Rather, they arise from changes in the *level* of expression of *existing* genes. In a nutshell, the information from training is not passed into the sequence of DNA bases contained within sperm and egg. So, the inheritance of Lamark's "acquired characteristics" became a heresy.

Even before changes in the DNA sequences of genes were identified as the key to evolution, the inheritance of acquired characteristics was opposed by influential figures. One was Charles Lyell,[10,11] the geologist whose writings greatly influenced Charles Darwin and who, after the voyage of the Beagle, became one of his confidants. He ridiculed Lamarckian views: disbelief that the giraffe might have attained its long neck through a stretching of the parental structure then passed on to offspring. Lyell's view was not based upon evidence, yet it turned out to be correct. The giraffe is a long-necked creature because variations in the sequences of DNA bases of its ancestors' genomes allowed the selection of a creature with a unique ability to exploit the high canopy of trees.

But in the 1920s, Jean-Baptiste's views appeared to be supported by results from the laboratory of Paul Kammerer – born in Vienna in 1880. Kammerer was led to conclude that hard pads, which developed on the skin of toads following prolonged contact of the skin with stone, were then found in the next generation. But the data were falsified, for Kammerer was misled by an assistant who believed the proposal of his boss must be correct.[12] This deceit had tragic consequences for, only six weeks after it was exposed, Kammerer committed suicide.

Yet Jean-Baptiste, born in the village of Bazentin-le-Petit in the north of France, was a great zoologist. Charles Darwin wrote of him in 1861: "He first did the eminent service of arousing attention to the probability of all changes in the organic, as well as in the inorganic world, being the result of law, and not of miraculous interposition." And, perhaps appropriately, acquired characteristics are now known to pass through at least a few generations. Mice given a diet rich in folic acid – a substance acting to add methyl groups to the DNA base **C** – produced visibly changed offspring, including a darker colour of coat. Not due to changes in sequence of inherited DNA; instead due to *a reduced level of expression –* surprisingly, heritable – of a single gene.[13]

Among diverse organisms, changes in the level of gene expression are heritable for more than one generation;[14] and other data, whose interpretation would otherwise be more hazardous, are in accord with these results. Towards the end of World War II, the severe shortage of food in parts of Holland caused tens of thousands to die of starvation. Not surprisingly, the prenatal effects of this famine resulted in more common disease in children who were *in utero* at this time. But one group of the famine cohort were reported to produce *grandchildren* who were smaller than normal.[15] Change in forms and functions, due to changed levels of expression of extant genes, may be heritable for even many generations.[16]

It is important not to get carried away by the inheritance of acquired characteristics. When organisms are, in following generations, modified in their forms through the influence of culture,[17] it overwhelmingly happens because changes in the sequences of DNA bases have been selected. Through these DNA sequences, life is stuffed full of information: the letters of the bases; the words of the triplets; the long sentences of the genes; the books of the chromosomes; the libraries of the genomes. Many dusty volumes are not opened; occasionally, some sentences read more or less often in offspring as a result of the experience of a preceding generation. But in all inheritance, the transmitted information is vast. How has this information, which produces awesome functions, been selected?

5

Darwin's Agents

The Premises of Darwinism

Ignorance more frequently begets confidence than does knowledge: it is those who know little, not those who know much, who so positively assert that this or that problem will never be solved by science.

– Charles Darwin

Molecules are the creators of life and, within organisms, have become complex. How, in the journey of matter, was this complexity selected? Charles Darwin (1809-1882) could not address this question at the level of these agents, for too little was known about them during his lifetime. But he did address it at the level of organisms.[1-4] And it is the premises of Darwinism that allow an understanding of the molecular complexity of organisms,[5-6] and how astonishing devices were selected.

Darwinian selection is based upon three premises. First, that there is variation. Second, that among the variants, there are differences in the fitness for survival. Third, the fittest systems are selected for survival in preference to the forms that function less efficiently – and, of course, these differences in fitness must be heritable.

Darwin published *The Origin of Species* in 1859. He was aware that his theory could cause such a furore that it might be better published after his death rather than before. Perhaps controversy would also cause distress to his wife Emma (a member of the influential Wedgwood family of pottery fame), who was a devout Christian. But Darwin's hand was forced when Alfred Russel Wallace (1823-1913) reached similar conclusions during travels in Malaysia and Indonesia. When controversy did arise, it was Thomas Huxley (1825-1895) (often described as Darwin's bulldog) who carried the flag for evolution – with its first requirement of variation.

Variation

Managers of sports teams use variation in players in the hope that, in the long run, they will be able to select a better team – a more successful combination of players. Nature has used variation at the level of molecules. It varies bases of DNA instead of players; its teams are genes and combinations of genes. And it generates its variation through mutations – processes through which the sequence of DNA bases in genes, or the amount of genetic information, is changed.

Mutations occur when light, or other forms of radiation, changes the structure of the DNA. Such damage may cause a lack of control in the division of specific cells – melanomas, or the cancers following the Chernobyl nuclear disaster. Chemicals from the environment may also cause mutations of DNA – benzene exposure possibly leading to cancers; "agent orange", used during the Vietnam War in the second half of the twentieth century, inducing malformation of the foetus. All these are variations with "bad news" outcomes.

Mutations also occur when mistakes are made in copying DNA sequences; these find analogies with the errors made in written text.[5] The most common mutation is the *substitution* of one base pair for another (cf "the newspapers have exorcised sound judgement"). In other cases, one or more base pairs may be *inserted* (cf "viewed through an ultra-violent lens"), *deleted* (cf "I can speak just as good nglish as you"), or *inverted* (cf "whilst we will provide tea and coffee during the morning and afternoon breaks, lunch will not be provided although you are of course welcome to sue the dining room." To produce variation, football managers are allowed substitutions and inversions; insertions are forbidden; deletions forced through a showing of the red card.

If errors in copying were too frequent then organisms would soon suffer fatal consequences. How is an adequate level of fidelity ensured in copying – fidelity that will normally lead to quality products?

> *The message 'Send reinforcements, we are going to advance' was*
> *sent, by a military man, through many intermediaries back to HQ.*
> *It arrived as 'Send three and fourpence, we are going to a dance'.*
>
> – UK anecdote

Attempts to transmit information are often thwarted by a lack of fidelity in the copying process. As the number of copying steps is increased, the information loss can become catastrophic, with loss of the original message. Evolution, based on the almost endless copying of DNA sequences, had to crack this problem. When DNA is copied, it can be estimated that, in the absence of special tricks to reduce copying errors, an error will be made about once for every 10,000 bases copied. This error rate is intolerable if an organism such as a human is to survive.

The problem was solved by the evolution of repair systems (enzymes) that, during the copying process, check the copied strand of DNA for fidelity. In copying, the **A** base in an existing strand normally ensures that **T** is installed in a new DNA strand, through the recognition of **T** by **A**. Rarely, an error is made and a base other than **T** may be installed opposite **A**. At the point at which this error occurs, there will be a "lump" within the new DNA double helix generated through copying. The "lump" arises because two bases, which do not fit together perfectly, have been put opposite each other (in the otherwise complementary double helix). The enzyme recognises this lump, and the aberrant base is then cut out (by protein-catalysed chemical reactions) and replaced by the correct base. Surgery to remove lumps that threaten survival has been going on for hundreds of millions of years.

Yet, without changes in DNA sequences as a function of time, evolution cannot occur. Hence, the error rates seen today must have evolved to optimise survival chances and reproductive success of the end-product organisms. The changes in the DNA sequences allow the micro-components of organisms to evolve – not only in structure, but also in the timing and amount of their production.

How many micro-components are there? It appears likely that about 250-500 proteins are necessary to make the simplest extant organisms; humans have perhaps a 1,000 times as many. The production of around a quarter of a million devices to make a human may seem like not many. But they often work in combination, thus permitting greatly more functions and forms than when acting alone. But how are these functions and forms selected?

> *The statistics John Wesson has compiled in 'The Science of Soccer'*
> *show that Premiership football players are vastly more likely to*
> *have been born in the first half of the school year. These were the*
> *biggest boys in the class and were thus selected for the school team.*
> *How fair is that?*
>
> – Daniel Finkelstein, 'The Times', London

Daniel Finkelstein's question is easily answered – natural selection, whether of teams or organisms, is not about being fair. It's about survival of the best adapted.

The quotation poses an interesting point about fitness and selection in society. Why is it that the birth of a boy in the period September-February – clearly advantageous for soccer selection at school – is still advantageous in soccer performance in adult life? Once the initial team had been selected, the unit had its own inertia: those initially selected to the team could then improve their abilities through repeated performance.

But biological selection rewards only good performance that is heritable. How do genes for efficient function come to be selected? Obviously, in a world of random mutations, those that have an adverse effect – or no influence – upon the survival fitness of the organism are enormously more probable than any that might be beneficial. After all, tinker with a television set in a random manner and the outcome is more likely to make it worse than better. This is a major reason why many people find it difficult to accept Darwinian theory. But, relatively rarely, mutations are beneficial for survival. Selection among new genes permits the evolution of greater complexity and fitness for survival. Nature did not care that far greater numbers of packets of matter were sacrificed in the processes of variation and selection.

Darwin's proposals, now supported by a mountain of evidence, turned the whole problem of the existence of complex organisms completely on its head. Hitherto, it had been believed that the production of intelligence required a designer. Darwinism required only a relatively simple set of rules and blind chance. The effects of evolution are seen even on a short timescale – in the year by year appearance of new influenza viruses, and in the increasing drug resistance of HIV and bacterial pathogens. On a slightly longer timescale, they were seen in the common occurrence of dark pigmented moths in the United Kingdom in a coal-based economy. Birds predated on the dark moths less successfully when they were camouflaged on the surface of blackened trees. When the coal-based economy largely disappeared in the second half of the twentieth century, lighter pigmented moths then reappeared in larger numbers.

Over a few billion years of natural selection, awesomely complex living machines evolved. Some critics of the hypothesis of natural selection fall into a common trap. A well-known critical argument is couched in terms of the almost zero probability of a monkey, typing in a random manner, producing the works of Shakespeare. The analogy has no validity. It is improbable in the extreme that a monkey would, given a vast amount of typing effort, produce even the small Shakespearian construct: "IF MUSIC BE THE FOOD OF LOVE". But it is not extremely improbable that the monkey would, after a while, type "IF", or "MUS". But the probability that it would type these two combinations at the same attempt is very small.

But natural selection does not work like that. In an analogy to the evolution of DNA sequences that promote survival, should "IF" serve such a function, it would be retained. Similarly, if "MUS" also served a survival function, it could evolve at a different time, and would also be retained. In an analogous manner, "HE" might evolve to join "IF MUS", and later "FO" to produce "IF MUS" and "HE FO" within the same sequence. The analogy of the monkey typing at random is fatally flawed because, in contrast to random typing, natural selection retains, at each stage of the selection process, that which is most useful.

Great thinkers in one area of science are sometimes unable to grasp concepts crucial in another – the training and models often very different. A few decades ago in the national press of the United Kingdom, the famous physicist Fred Hoyle expressed the view that the evolution of a protein by chance mutations was to be likened to the spontaneous assembly of a Boeing 747 from components in a scrapyard. He had fallen into the classical trap.

The error was exposed in a letter from the then President of the Royal Society of London, Sir Andrew Huxley. He pointed out that the protein evolved because useful function was selected over a very large number of small stepwise selections. Hoyle should not be criticised for expressing an erroneous opinion; rather congratulated for bringing a matter of such importance forward for discussion. But let's be clear: Sir Andrew was right; Fred was wrong. And in selecting useful functions, sex opened up new opportunities.

...those things which increase passion should be done first, and those which are only for amusement or variety should be done afterwards.

– Kama Sutra, 1883

The great drive for sex is proved, in a least one species, by the tabloids: a passion that, if absent, is a death sentence for a package of about 20,000 genes. Amoebae – reproducing simply by the division of one cell into two – lack this passion. The two daughter cells are clones – each taking a copy of the genetic information that existed in the original cell.

However, mechanisms that allow more rapid variation may be a good thing, allowing the evolution of new machines better fitted for survival. A mechanism resulting in more variation was provided by the evolution of sex around one billion years ago. Consider random changes, and selection, again for generation of the Shakespearian text: "IF MUSIC BE THE FOOD OF LOVE". After a very large amount of random typing, and selection in the Darwinian manner, the advantageous "IF MUS" might evolve in one case; "HE FOO" in another. But, in the case of DNA sequences, to get these advantages to occur in the same string is, in the absence of sexual tricks, going to take a very long time.

During the formation of the sperm cells of Fred and of the egg cells of Mary (be they sea urchins, horses or humans), pairs of chromosomes line up – end to end. The position of the pairing is not random; rather the genes that code for products with the same function are paired. The lined-up DNAs are then cut, and the bits recombined. Through such a cut, the advantageous mutations "IF MUS" and "HE FOO", although on *different* genes prior to recombination, can end up on the *same* gene. And, should these sex cells give rise to offspring, then the advantages are combined in the offspring. Disadvantages may also be combined in this way; but then natural selection acts against these unfortunates. Sexual reproduction also gives rise to new combinations of genes for, if two genes that enhance survival probability exist one in a female and the other in a male, they can end up in the same package.

Observations of the vehicles for sexual reproduction – eggs and sperm – and of the bodies that are produced upon their combination, led to the famous conundrum: "Which came first, the chicken or the egg?" – a classic example of a question based on a false premise. The chicken did not precede the chicken egg, nor *vice versa*. Bodies, and the vehicles that transmit the genes required to make the bodies, co-evolved. In a sense, chickens and humans are guardians of sperm and eggs; dispensable once they have delivered their precious cargo. Sex comes at a high price for its benefits and joys also demand the unpleasantness of ageing, dying and death.

Ageing and Death

Our life is made by the death of others.

– Leonardo da Vinci

Last scene of all,
That ends this strange, eventful history,
Is second childishness and mere oblivion,
Sans teeth, sans eye, sans taste, sans everything.

– William Shakespeare in 'As You Like It'

Reproduction through sex, and death as the final consequence of ageing, are inevitable companions.[7] Sex is a good way of producing fitter machines. Old and less fit machines must be removed to make way for new models. *Ergo,* the scrap heap; *ergo* the driving force for generations, their decay, their eventual demise.

But the rate of decay is very variable – sometimes remarkably slow, for the bristlecone pine tree survives up to 4,000 years; sometimes remarkably quick, for the adult mayfly lives for only a few days. Dogs show the characteristics of old age – grey hairs on the muzzle, relative inactivity and perhaps a limping gait due to arthritis – from around nine to sixteen years of age. In contrast, a human has not at this age even reached its peak of performance.

These large differences in rates of decay indicate that it is not that the materials of bodies can last for only a specified number of years. Rather that the rate of decay has been programmed in evolution's attempts to optimise the survival fitness of a species: seductively described as Death by Design;[8] more accurately as death by genetic programme. Lest you are not impressed by the idea that genes can actively optimise our rate of decay, take on board the observation of the great R A Fisher.[9] He concluded that, if a mutation gives to an organism a 1% advantage over those that do not possess it, within about 100 generations that mutation is likely to be widespread.

The crucial role of *specific* genes in ageing is established. In a worm, a mutation in a single gene doubled the lifespan.[10] Progeria is the disease that causes humans to age at about seven times the normal rate. At age 11, a sufferer appears around 80. The tragedy is a consequence of a single mutation in a gene on chromosome 1.[11] During evolution, lifestyles and genes involved in rates of ageing have been mutually dictating through feedback. So a lack of parental care correlates with rapid post-reproductive ageing. If the reproductive organs of Pacific salmon are removed, they have a much longer average lifespan than is normal; mate, and they die shortly afterwards.[12] In contrast, after *Homo sapiens* pass the optimum child-bearing age, they age much more slowly – consistent with the extensive parental care necessary in a species where complex social skills promoted survival.

Senescence and death are of course also promoted by a bad lifestyle,[13] and, conversely, the "good life" can defer adverse changes. But wrinkles arrive roughly on cue, no matter how much face cream is applied; muscle weakening only postponed by visits to the gym. Life expectancy is largely dictated by the vast history of our evolution – genetically dominated, but fine-tuned by the environment within our lives. And average life expectancies in the environments of "the good old days" establishes such days as mythical.

Near to the end of his life, Frederic Chopin commented, "You and I are a couple of old cembalos on which time and circumstance have played out their miserable trills...In clumsy hands, we cannot give forth new sounds and we stifle within ourselves those things which no one will ever draw from us, and all for lack of a repairer." Chopin had it wrong. There are repairers, but there are also active destroyers, and the destroyers limited the time for him to compose his intensely moving music.

The molecules appear almost as magicians. But how do they know when, and in what quantities, to make their wondrous products?

Production – Quantity and Timing

'I liken it to a swan,' said Valerie. 'It breaks my heart to think it won't fly any more.'

– Comment of a French citizen upon the demise of Concorde

Concorde was an aeroplane with beautiful lines, evoking national pride in France and the UK; but very limited in passenger capacity, a voracious consumer of fuel, and noisy to boot. Reality came home to roost: production ceased.

Factory systems should produce efficiently and only that which has utility; and neither more nor less than is needed, and at the right time. The factory managers make conscious decisions. But even these decisions are the behaviours that are programmed into the collections of molecules that elicit them. How do molecules "know" what is needed and when?

It was French scientists who discovered the principle that allows quantity and timing control in organisms. Francois Jacob, Andre Lwoff and Jacques Monod showed how the expression of a gene, to give its functional product, is controlled by molecules whose very presence indicates that the gene product has a job to do. For their crucial discovery, they were awarded the Nobel Prize in Physiology or Medicine in 1965.

The system they investigated involved a gene product (an enzyme) needed to utilise a certain food molecule. They found that the enzyme is produced when the food molecule is around. Conversely, when the food molecule is not around, the translation of the sequence of DNA that codes for the enzyme is switched off. This switching off is achieved by the action of a blocking molecule – a repressor – that recognises and binds to one end of the gene. The expression of the gene is prevented, and the enzyme required to utilise the food molecule is not produced.

However, in the presence of the particular food, a product from the food binds to the repressor molecule. In so doing, the product from the food changes the shape of the repressor molecule so that it no longer has a high affinity for the DNA. As a consequence, the repressor now drops off the DNA. The DNA can now be expressed to give the enzyme required to metabolise the food.

This system, generating what is needed to metabolise the food only when the food is available, gives a fundamental insight into how life's processes are controlled by molecules. Natural selection allowed molecules to achieve quantity control. This "feedback control" is analogous to the way that factory products are controlled according to the laws of supply and demand – production lines suppressed when their products are not needed during a recession.

It is a salutary thought that automated molecular systems have reached such complexity that consciousness has evolved. Consciousness, and indeed all structure and function, is operated by automated systems of molecules that lie beyond your control. They are controlled according to programming. We are approaching the heart of the matter. But hearts, brains and bodies are much more than micro-devices. How are these more complex objects forged?

Building Complex Objects

One contract is contingent on another in terms of time and schedule. So, with 365 general contractors and thousands of subcontractors, if one got off schedule so would the others.

– Ginny Greiman, risk manager of Boston's Central Artery/Tunnel project with a budget (as of 2001) of US $14 billion

Complex biological machines are products of the evolved and coordinated actions of enormous numbers of micro-devices. These machines display astonishing powers because "the whole is greater than the sum of the parts". This property is sometimes used to challenge the scientific approach as "reductionist" – to argue that a full understanding of complex organisms in terms of an analysis of their components is not possible. This is a truism, but science does not work this way. It works through an approach that was beautifully put by the molecular biologist Sydney Brenner: "The property of the whole is equal to the sum of the parts *plus the interactions between them.*" To fully understand organisms, we require an understanding not only of the components, but also of all the interactions between them. And the number of interactions is astronomical – which is why progress is slow.

In making complex products, there are advantages in producing all the bits and pieces in one location under the control of only a few managers. The many parts, or components, must be produced in the appropriate quantity and at the appropriate time, as in Boston's Central Artery/Tunnel project. In order that biological machines could evolve, molecules had to develop production management of myriad other molecules. Single molecular signals·– central managers – evolved for the switching on of *large numbers* of genes, in a manner analogous to the switching on of a production line at 8 am. The protein products from the translation of these genes then assemble the complex object.

An example of impressive complexity – but still simple compared to the switching on of cascades of genes that result in the building of animal bodies – is found in the construction of molecules used by microbes and fungi to fend off their competitors. Such molecules are not esoteric from the viewpoint of humans – they are antibiotics. These antibiotics are awesome in their power to kill the microbes that infect, and often kill, humans. The so-called humble microbes have defences that, in some circumstances, are superior to those manufactured and employed by the human body.

For the discovery of penicillin, and for the realisation of its use against otherwise lethal human infections, Alexander Fleming, Ernst Chain and Howard Florey were awarded the Nobel Prize in Physiology or Medicine in 1945.[14] Although penicillin was initially successful against a very wide variety of bacterial infections that threatened human life, these pathogens quickly evolved resistance to it – through Darwinian selection – as it came to be widely used. An end product of this resistance is the super-bug MRSA (**m**ethicillin-**r**esistant *Staphylococcus aureus),* responsible for hundreds of thousands of human deaths per annum in the 1990s. As resistance became an increasingly serious problem in the period 1960-2000, pharmaceutical companies searched for new and more efficient antibiotics.

Amongst the new discoveries, vancomycin became the most famous "antibiotic of last resort". As is typical for antibiotic construction, the microbe that produces vancomycin uses a production line, which involves products from about 30 genes;[15] a production line switched on by its managers only when needed. The line requires not only the materials of which the antibiotic is made – 7 distinct and rather unusual building blocks – but it also requires the tools – enzymes – to fasten these building blocks together. The manufacturing process takes place by joining of the 7 units (numbered 1 to 7) to give successively larger constructs: 1-2, 1-2-3, 1-2-3-4…and finally 1-2-3-4-5-6-7. But the sophistication of the production line does not end there, for additional connections between the building blocks must be made, eg 2-4, 4-6, 5-7; and sugars added to ensure water solubility, and simultaneously improve the efficacy of the final antibiotic product.

That final product is in the shape of a bowl, but less than a millionth of a centimetre long. The bowl "grabs" a part of a growing microbe and, like the mediaeval bulldog selected by humans to hang on to bulls, sticks to its target. But this is an evolved cohesion for which we can all be grateful; pathogenic microbes die, humans live on. Should you see a picture of a microbe, look at it with new respect – for the products of microbes are used to kill microbes.

The antibiotic production line, even if simple compared to that required for the production of bodies, is extraordinary in its beauty and economy. It illustrates the incredible power of natural selection; how, in the construction of complex systems, cascades of genes are switched on by one, or a few, control elements – knock over one domino and all the others follow. The necessity of central management of complex projects was discovered by molecules hundreds of millions of years before applied in human mega-projects. Human discoveries can be regarded as evolution accelerated by the evolution of mind.

Contiguous clusters of genes, under the control of a single switch, are unsurprisingly also found in animals. But surprisingly, the many genes may also be peppered about on different chromosomes, as opposed to being located at one site. This begs the question: "How are they then all switched on at the appropriate time?"

In animals, the DNA memory is retained inside a bag (nucleus) at the core of the cell. But all the proteins derived from it are, through intermediate RNA, produced outside this bag – in the main bulk of the cell. Master switches – proteins known as transcription factors – must therefore be generated in the main bulk of the cell. They are then transported back into the nuclear bag. It is there that they are able, through diffusion, to "search around" and switch on synthesis of all the required proteins – on whichever of the chromosomes their cognate DNA lies. The production manager walks around the factory telling many satellite sites that production should begin now.

"A gene for this, and a gene for that", much loved by the press, can be a dangerous oversimplification. For some of the complex constructs necessary to make the exquisite machinery of animals contain more than a hundred protein components. How did evolution lead to a very large numbers of genes with a variety of functions?

Modifying

History is a gallery of pictures in which there are few originals and many copies.

– Alexis de Tocqueville

The copying of successful devices is common – almost a way of life. My abiding experience of this phenomenon occurred when being shown round a factory in Japan in the 1970s. An instrument, a recent and successful product manufactured in another country, was being painstakingly dissembled. "What is this?" I enquired. "Oh," came the embarrassed reply, "competitor's product." Something successful in one place is copied by others; sometimes then modified to give improved function or even modified to provide a new function. Genes have been at the game of copying and modifying from time immemorial.

With the passage of time, organisms occasionally acquire – by a well-established mechanism during recombination – an extra copy of a gene. A great advantage in possessing two copies of a gene is that one copy can evolve while the original function is still provided by the original structure. "Tinkering" can occur with a lower probability of being harmful. This is a very good way for possibly increasing survival fitness since, in some cases, the evolving structure takes on a useful new function. The process of gene duplication, followed by modification, has been a common pathway for the evolution of genes with new functions.

Hijacking Genes

If there is something to steal, I steal it!

– Pablo Picasso

Vertical gene transfer is the process occurring when copying of the genome gives rise to a new generation of an organism. The worldly goods of an organism have been bequeathed to its own. But, as in human affairs, outsiders would often like to get into the act.

Stealing your neighbour's goodies can be a good survival strategy. The goodies taken from neighbours are genes. To differentiate this inter-species transfer from the normal processes of heredity, it is called *horizontal* gene transfer. It is, for example, the means by which antibiotic resistance is transferred from one bacterium to another, resulting in the widespread problem of MRSA – bad for the survival of humans; good for the survival of bacteria. The occurrence of this "gene smuggling" is becoming increasingly recognised.

Horizontal gene transfer could account for some astonishing observations. A plant (a *solanum* species) growing in Argentina was investigated because cattle eating it suffer from a wasting disease characterised by abnormally high concentrations of calcium in the blood. The plant was discovered to produce a sugar derivative of the human hormone that initiates the absorption of calcium into the body – without which humans are unable to make their bones.[16,17] This hormone does the same job in a wide variety of birds and mammals, and only a few millionths of a gram are required per day to produce and maintain human bones. Since ingestion of the plant by cattle provides massively larger amounts of the hormone derivative, they become overloaded with internal calcium.

But greater astonishment comes with the knowledge that the hormone is made in the mammals by a tortuous path. The hormone precursor is vitamin D – a molecule that is structurally related to cholesterol. The vitamin D – provided in the diet or, alternatively, when sunlight falls on the skin – first passes to the liver where a complex protein causes an oxygen atom to be inserted into its structure at a specific point. This product (25-hydroxyvitamin D) next passes to the kidney where another complex protein causes it to add a second oxygen atom – at a point remote from the first oxygen addition (to produce 1,25-dihydroxyvitamin D). So, two complex proteins, one produced in mammalian liver and the other in kidney, are required to elaborate a vitamin to produce a hormone.

Did a plant acquire from a mammal the genes needed to make these proteins – and thereby acquire the advantage to deter browsers? Or did mammals acquire these genes from a plant? Or could evolution be so convergent that the requisite genes evolved independently in the plant and in mammals? Or is there a common ancestor of plants and animals that possessed the prototypes of these genes? These questions need to be addressed, for if horizontal gene transfer is responsible for the observations then it has wider implications than hitherto realised.

Horizontal gene transfer is raising its head in other important areas. For example, one species of animal (a rotifer, an animal a few millimetres in length that lives in lakes) manages to survive without sex, reproducing itself instead by simple cell division (asexual reproduction). This is very surprising because rotifers used sexual reproduction in their earlier history, but later reverted to asexual reproduction. The famous, and supremely engaging, evolutionary biologist John Maynard Smith described the rotifer as an "evolutionary scandal";[18] for if you have something that promotes variation and survival fitness, why dump it? There is now a hint that these rotifers continue to thrive because they benefit from horizontal gene transfer.[19] They have garnered genes from bacteria, fungi or even plants; hijacked their tools for survival from others.

Had such processes been the rule, rather than the exception, the picture of the tree of life would have been discarded. Organisms would not have a well-defined lineage. Instead, they would have been a composite made up to include bits of relatives – even very distant ones. However, the available genome sequences of animals suggest a rather watertight history, although some unanticipated complexity may surface in the future. The tree of life looks secure – albeit that some of its branches have been cross-linked *via* the brushing together of their twigs.

I only know what I read in the papers.

– Will Rogers, American journalist

Single base mutations, sexual production and gene duplication are ways in which, given selection, the survival fitness of an organism can be improved by the generation of new genes. And jumping DNA promotes the evolution of complexity. But we need to know the sequence of bases in the evolved genomes since this knowledge will increase our understanding of how living organisms have evolved, are assembled and function. In the 1970s, Fred Sanger, working in Cambridge, UK, developed an astonishingly rapid, sensitive and elegant method to read the books of life.

It was exciting to be around in Cambridge at that time. It was also instructive in illustrating how, as throughout history, few see what those disposed towards more careful thought are able to see. For the common perception was that DNA was "a boring polymer". What sensible person would be interested in long sequence of only four bases repeated almost *ad nauseam*? What sensible person would have the confidence to believe the massive numbers of sequences *could* be determined? Only those who had a great deal of self-belief, could see a big problem, and were wise enough to realise that the information contained therein determined the nature of organisms – and a good deal of their behaviour. Fred Sanger showed all these qualities.

Where a portion of the sequence of a genome is -G-A-A-T-C-, the automated copying process of Nature produces the complementary strand -C-T-T-A-G-. In Sanger's method, four parallel copying experiments are carried out. In all four experiments, the copying machinery is provided with normal building blocks of G, T, A and C. But in Experiment 1, a trace of a modified form of C that does not have the "hook" ("hook-less" C) required to continue the growth of the chain of DNA is added. Experiments 2, 3 and 4 respectively contain some "hook-less" T, "hook-less" A and "hook-less" G. Where a base lacks the "hook" to continue the growth of the copied strand of DNA then that DNA chain must terminate its growth at that point. Thus, since the chain is copied in the direction right to left, the outcome of the four experiments (copied strands shown in italics below template strands) is:

Experiment 1

     –G–A–A–T–C   ⟶   –G–A–A–T–C
                                       –*C*–*T*–*T*–*A*–*G*

Experiment 2

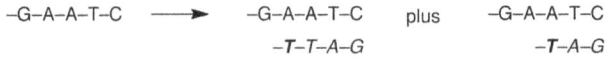

     –G–A–A–T–C   ⟶   –G–A–A–T–C   plus   –G–A–A–T–C
                                    –*T*–*T*–*A*–*G*   –*T*–*A*–*G*

Experiment 3

     –G–A–A–T–C   ⟶   –G–A–A–T–C
                                       –*A*–*G*

Experiment 3

     –G–A–A–T–C   ⟶   –G–A–A–T–C
                                       –*G*

Figure 5-1. The four experiments of Sanger sequencing.

The "hook-less" bases are shown in bold font and must always be found at the end of a growing chain since they lack the chemical feature that is necessary for further growth of the chain. The "hook-less" bases are the only radioactive bases present in the experiments, and this radioactivity allows the detection of the terminated strands (*and no others*) with extremely high sensitivity. The relative lengths of the copied strands govern the extent to which they travel on a solid surface – the shortest chains travel furthest from their point of application to the surface. Thus, as a lengthening copied chain is identified in the sequence of experiments 4 > 3 > 2 > 1, the experiments determine that the bases are added in the order G, A, T, T and C.

Such is the sensitivity of the detection of the radioactive "hook-less" bases that DNA sequences could be determined using less than a thousandth of a millionth of the gram of the template DNA. Given this technology, the sequence of many tens of thousands of bases (in any selected genome) could be determined in a day. For invention of the method, Fred Sanger received his second Nobel Prize in 1980 (shared with Walter Gilbert). His first Nobel Prize (in 1958) was for the first determination of the sequence of amino acids in a protein. Fred was renowned for his modesty. When the Laboratory for Molecular Biology celebrated the award of his second Nobel Prize, successive speakers emphasised this quality. Finally forced to his feet, he teased: "Well, if I've won it twice, I must be bloody good!"

By 2003, the above method for the sequence determination of DNA, given some advances to increase the efficiency of the technology, was used to sequence – with the exception of a few details – the human genome. The cost was of the order of a billion dollars. The sequence contains the information that instructs the development of a fertilised egg, when bathed in the environment of a mother's reproductive system, into a baby; information that, given its earthly environment, will allow it to function during its lifetime.

By 2009, even smarter and faster methods for gene sequencing have been invented. Cheaper determinations – by a factor of about 10,000 – of the instructing codes of life have arrived.

Copying Packaged Libraries

The four DNA bases have been likened to the letters of a language, genes to its sentences, chromosomes to books, and genomes to libraries. For life to continue to exist, the libraries must be copied. Although the double helix solved the problem of copying, the libraries of life are, unlike those of true libraries, tightly packed in the convoluted structures of chromosomes – structures more akin to balls of wool than neatly ordered pages and volumes. How is the information on the woollen thread accessed?

Our DNA is contained in the nucleus of each cell (excepting red blood cells), and separated into 46 "balls of wool" (46 chromosomes). Each chromosome is a single molecule of DNA and, when stretched out, has an average length of about five centimetres. Such is the number of cells in a human that the combined length of the DNA in a human is sufficient to stretch around ten times from the Earth to the Sun and back.

In each cell, the chromosomes are present as duplicate copies, so there are actually 23 pairs. An advantage in carrying a second copy of the genetic information is that, if one copy contains adverse mutations, a normally functioning copy may remain. The egg and sperm cells – containing only one copy of each of the 23 chromosomes – are exceptional. Thus, when egg and sperm cells combine in the first step towards making a new individual, the norm of 46 pairs of chromosomes per cell is restored. Every time a cell divides, the whole of the 3,000,000,000 base pairs in the chromosomes is copied. The long strings of DNA – the 3×10^9 bases that are twisted around a complementary sequence of the same length in the double helix – must be unwound to do the copying.

It is as if two ropes of 1-inch diameter were coiled one about the other so that the length of the coils was about 40,000 miles, and this rope was packed into a hollow building about 200 feet high, 200 feet long and 200 feet wide. Every single inch of the rope is colour-coded red, blue, yellow or green – corresponding to the four bases. The coils must be unwound and copied, and each new coil wound round one of the old strands and refolded. If mistakes are made in copying the 3,000 million bands of colour, they must be automatically corrected with high fidelity. If the molecular equivalent of this operation were not carried out quickly, automatically and with high fidelity, we would soon be dead – for it is executed countless billions of times in our bodies every day.

The realities of Darwinian evolution were, as noted by Steve Jones,[20] not appealing to George Bernard Shaw: "When its whole significance dawns on you, your heart sinks into a heap of sand within you. There is a hideous fatalism about it, a ghastly and damnable reduction of beauty and intelligence, of strength and purpose, of honour and aspiration" – perhaps why natural selection has been, and still is, so often opposed. The evolution of bodies, one of natural selection's most awesome outcomes, is considered next.

6

From Bits To Bodies

The Evolution of Organisms

> *We can allow satellites, planets, suns, universe, nay whole systems of universe[s] to be governed by laws, but the smallest insect, we wish to be created at once by special act.*
>
> – Charles Darwin

Today there is an astonishing variety of complex bodies.[1] In the determination of the times at which the prototypes of these bodies appeared, two methods are particularly important.

Only one of them – radioactive decay – measures time in absolute terms. The slow decay of radioactive potassium to argon dates very ancient events (chapter 2); the faster carbon to nitrogen conversion dates events that occurred only a few thousand years ago. Since rocks are dated in this way, the age of the fossilised species found in them are also known. Additionally, at the surface of the Earth, the normal expectation is that older rocks tend to lie under younger rocks. Therefore, fossils that are found in low-lying strata can, with a few exceptions due to unusual upheavals, be concluded to be more ancient than the ones found in the upper strata.

A second "clock" is provided through variations in DNA sequences. It gives the greatest detail for inferring the times at which specific species appeared on Earth. The basic idea is that mutations may occur in DNA at a rate that is constant. If this were so then the differences between two organisms in terms of their DNA sequences could be used to measure the time that had passed since they had a common ancestor. However, reality is not so simple.

Some DNA that appeared early in evolution has, over eons, reached an optimised structure. Mutations in such DNA are quite likely to be lethal to the owner. So, such DNA sequences barely change with the passage of time. This is true of the DNA that codes for the histone proteins that "package" DNA in the chromosomes. These histones are essentially the same in oak trees, in peas and in humans. They bear analogy to the sloped roofs of buildings – long tested and proved; mutating them to a version that is flat and the outcome may be a disaster. In contrast, a high mutation rate is advantageous in some DNA sequences, such as those coding for the proteins of the immune system. In these cases, variation is a good thing since it allows the generation of an enormous variety of structures – used to recognise a plethora of foreign invaders.

So, mutations appear at different rates according to the roles of the gene products. But there are circumstances in which DNA sequences can be mutated without consequence for the organism. One circumstance is where the mutation changes the DNA sequence but nevertheless – because of degeneracy in the genetic code – gives rise to the same amino acid in the protein. In this situation, the mutations are "neutral mutations", and as such should more accurately reflect an "intrinsic rate". Given that only neutral mutations should be considered, but with the caveat that the repair rates of mutations can vary from organism to organism, changes in DNA sequence give an astonishing wealth of information on the evolution of species.

Through the application of these dating methods, a lot is known about what came from what, and roughly when – detailed trees of life showing when distinct species arose from a common ancestor. The evidence suggests that all organisms evolved from a single single-cell. Strikingly, the devices required for the copying of DNA, for the conversion of RNA into protein and for the conversion of RNA into protein are similar in all organisms. Last, but not least, of the two possible mirror image forms of the amino acids, protein assembly involves one form in all organisms. The single tree of life is not fanciful but based on hard data.

Life's Variety

Man still bears in his bodily frame the indelible stamp of his lowly origin.

– Charles Darwin

Variation and selection in different environments allows differing entities to evolve from a common ancestor. If geographical features isolate the descendants of a common ancestor for a long period then the evolution of new species can occur. Richard Dawkins[2] has given an account of the evolution of the main animal species – what evolved from what, and when. In 1978, Shorrocks[3] presented estimates of, among others, about 41,000 species of vertebrates and about 900,000 species of arthropods – although it was commented that the latter number could be as high as several million, reflecting the enormous diversity of insects. He also estimated around 100,000 species of molluscs and nearly 300,000 species of flowering plants. But such is the rate of change of views, and so enormous is the task of identification, that the current view is that most species of life have yet to be uncovered, with the total number estimated possibly as high as several tens of millions.

Even more uncertain than the number of living species is the number of species that have ever existed; estimates range from about 1 billion[4] to even hundreds of billions. The large variations indicate that we just do not know, although there is a consensus that the vast majority of them met their demise. This consensus allows paleobiologists to speculate on the average lifetime of a species – from around 100,000 to a million years. If these estimates are to be believed then modern humans (*Homo sapiens*), having already been around for about 200,000 years, is doing reasonably well. But in the numbers game, we are still pretty small beer. Single colonies of ants may contain in the region of 20,000,000 individuals; and in a typical pinch of soil, microbes are found by the million.

Forging Humanity

A few forms of life are given in Table 6-1. "Appearance" means the time at which the organisms became unable to breed with their immediate common ancestor. Because of uncertainties in the fossil record or of dating methods, the times of appearance are approximate.

Table 6-1
Rough Timescales for the Appearance of Various Forms of Life

Type of Organism	Appearance (years ago)
First Cells	3.5 – 4 billion
Bacteria	2 – 3 billion
Photosynthetic bacteria	*ca* 2 billion
Green plants	*ca* 1 billion
Fish with bones	500 million
Insects	400 million
Land vertebrates	350 million
Dinosaurs	300 million
Birds	250 million
Mammals	110 million
Ape-like primates	40 million
Hominids★	*ca* 6 million

★*This class includes not only Homo sapiens, but also the species that preceded them following the bifurcation that also led to modern chimpanzees.*

Table 6-1 misleads in two important respects. First, it tempts us to believe that the appearing forms would be similar to those of today – true to a degree that makes the table useful; false insofar as all extant forms have gradually changed from their prototypes. When the lines leading to the chimps and humans diverged some six million years ago, their last common ancestor had features strikingly similar to both species of today; but was neither in detail. Second, the table gives a false impression of a staircase of evolution with humans at the top. In truth, humans are a miniscule part of the species game; the tree of evolution possessing myriad branches with humanity as but a tiny twig. Through variation and selection, these branches sometimes burgeon; are sometimes pruned; occasionally cut off altogether.

Insects which do not – and cannot – grow very large by the average mammalian standard retain their shape by means of a surface structure – an exoskeleton, composed of chitin. As larger organisms evolved, to avoid gravitational warping, so did bony skeletons. The calcium within bone has lost some of its electrons. So, in bone there is a stronger expression of the electromagnetic force than in chitin, and as a consequence bone is less plastic than chitin. Even bone has its limits in opposing the effects of gravity; but a whale can comfortably be bigger than an elephant because of the upward push it receives through its displacement of water in an ocean environment.

The rates of change of biological machines are very variable. The theory of natural selection suggests that organisms will change little once they have evolved to be close to "as good as they can get" in an essentially fixed environment – sharks and many bacteria come to mind. Conversely, organisms that migrate into, and themselves forge, new environments change at a more rapid rate. The changes in humans, relative to the common ancestor shared with chimpanzees only about six million years ago, puts them in this group. Changing environments put evolutionary pressure for changes in both outward form and internal structures. Such changes, in turn, allowed new environments to be explored. This feedback, between changing environments and forms, promotes change that is relatively rapid. Interestingly, our whole picture of the evolution of humans in the period of pre-history (over a period of about five million years) is based upon no more than a few thousand skeletons – and most commonly very incomplete ones.[5] Sufficient to outline the story, but incomplete to a degree that leaves areas of ignorance.

Since our brains deal rather poorly with large numbers, the relative timescales for evolution of some features of the Universe and a few forms of life, normalised to the passage in total of one year, are instructive (Table

6-2). The Universe is so old that all of recorded human history occupies less than the last minute of the year.

Table 6-2
Rough Relative Timescales, Normalised to the Passage of One Year, for the Evolution of the Universe and Various Forms of Life

January 1: Occurrence of the "Big Bang"

Early February: Formation of the Milky Way

Mid-August: Formation of the Earth and the Sun

Late September: Appearance of first life on the Earth

December 13: Appearance of the first animals

December 25: Dinosaurs walk the Earth

December 30, midday: Dinosaurs meet their demise

December 31, noon: Lines leading to humans and chimpanzees separate

December 31, 11:59 pm + 30 seconds: Development of agriculture

December 31, 11:59 pm + 45 seconds: Building of the pyramids

December 31, 11:59 pm + 59 seconds: Newton formulates his Laws of Motion

As a consequence of the continental drift that was to follow the first appearance of bones (about 500 million years ago), the United Kingdom has moved from its then position in the southern hemisphere to its present northerly location on the globe. When the first animals and dinosaurs appeared, this land mass had moved to about 30° North.

Details of the evolutionary path to *Homo sapiens* have, not surprisingly, a special interest.[2] By about 1.3 billion years ago, there existed a single cell organism with its DNA inside an internal compartment (a eukaryote). This organism had itself taken about 2.5 billion years to evolve from the earliest form of life. Given a further 0.5 billion years, one highway from this eukaryote led to a ball of cells that used cilia (hair-like projections from the main body of the cells) in feeding and movement, and reproduced sexually. Given about a further 200 million years, one road from such multi-cellular agglomerates led to a worm-like creature with primitive eyes that lived on the seabed. It took a further 180 million years for one street from this worm-like creature to produce a fish-like animal, but one with substantial tissue in four lower-side oriented fins. So, by about 420 million years ago, the pods that would permit a migration of an animal from sea to land were appearing.

Among many selections from the fish-like animal, one produced creatures with features akin to the salamanders (after 80 million years). And from there, after a further 40 million years, one lane led to animals with features akin to those of today's lizards. From the latter, one path produced a nocturnal shrew-like animal in a further 200 million years. A track from the shrew-like animal to a large quadruped ape then took a mere 90 million years (although perhaps, more impressively, still 45,000 successive periods each as long as the whole of the 2,000 year Christian Era). Among the tracks from this ape, within eight million years, one led to an animal loosely recognisable by reference to today's chimpanzees. Six million years later, *Homo sapiens* – to be viewed as organisms at one point in evolutionary space among millions of species – walked the planet.

Complex organisms were not created as such – they evolved. The organisms of today have managed to survive through billions of years of harsh competition, with the consequence that they are awesomely sophisticated machines.[6] Their genomes have evolved to promote their survival, with the implication that they will be very tricky customers. In broad outline, science has answered the questions of the philosopher Søren Kierkegaard's *Young Man*[7]: "How did I get into the world? Why was I not asked about it? Why was I not informed of the rules and regulations?" And there is even much knowledge about the rules and regulations that gave rise to human variety.

Human Variation

Within the last mere century, through accelerating global transportation, migration and communication, the genetic and cultural variation of *Homo sapiens* is, with increasing rapidity, being assigned to the blender. But much can be learned from the way in which humans came to vary.

The absence of land carnivores in Antarctica aided the evolution of the flightless penguin. Variation among humans was determined by the same principles. Climate, geography, flora, fauna and minerals imposed the details of the forms of humans – the variety of the races – and their cultures. We are left to observe the successes – from adaptation in the cold Arctic wilderness to the parched deserts; and have lost knowledge of many of the failures. Through the evolution of agriculture, some groups escaped the almost eternal drudgery of hunter/gatherer and nomadic life. The abolition of drudgery facilitated the development of writing, mathematics, science, technology and political centralisation.

How did humans travel into different environments? Jared Diamond has summarised patterns of human migration over about the last million years.[8] About six million years ago, the species *Homo* diverged in Africa from an ancestor that was common with that of modern chimpanzees.

Its first migration from Africa probably commenced about a million years ago (Figure 6-1). But the line giving rise to modern humans (*Homo sapiens*) arose around only 200,000 years ago, with its migration from Africa occurring after about a further 100,000 years. This species, arising from a relatively small group, took over from the earlier migrants.

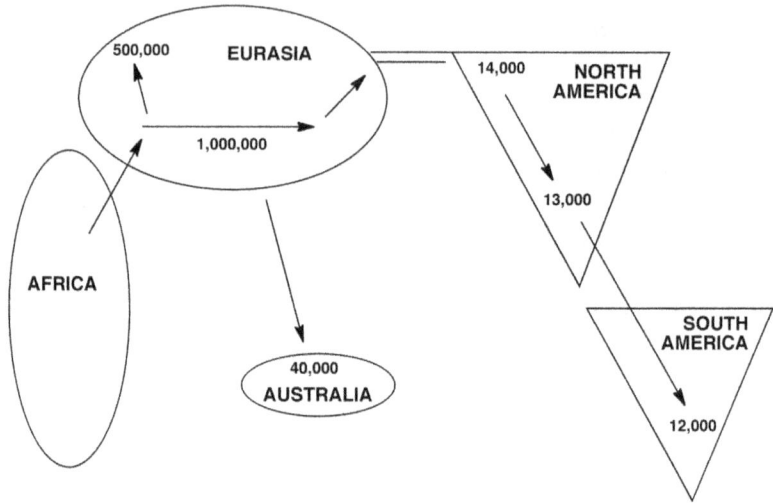

Figure 6-1. The pattern of human migration. The numbers give the estimated times (years ago, based on the fossil record, and in some cases also radiocarbon dating) at which the first migrations occurred (modified from Jared Diamond[8]). (Note that the migrations of modern humans from Africa occurred only about 70,000-100,000 years ago, and therefore their migration across Eurasia is much more rapid than conveyed in the figure.)

The fact that about fourteen billion years after the chaos of the early Universe, extremely complex packages of matter were journeying in forms that would invade almost the whole of the planet's land surface is astonishing, but was not unique. Other organisms – for example, plants and the bacteria that live in soil – had become widespread at much earlier times. But what was remarkable about human migration was the way the already enlarged brain of *Homo sapiens* was induced to increase further in size as it experienced new environments and lived in increasingly complex social groups. There was evolution of a being that could contemplate its existence.

The migrations, and the times at which they occurred, followed patterns sensible in terms of the relative positioning of the continents, and the presence, or lack, of bridges between them. The connection between Africa and Eurasia allowed early migration from the former to the latter.

But, following the exodus of modern humans from Africa, it took around 50,000 years for the Americas to be populated by migration across the shallow Bering Strait, very probably because of the inclement climate of north-eastern Asia. Once this migration had occurred, the spread of humans to the southern part of South America occurred relatively rapidly (within 2,000 years) – perhaps facilitated through migration *via* the coastal route of the American continents.

The colonisation of Australia, 40,000 years ago, had unsurprisingly taken a relatively long time, since progression across relatively large tracts of water – leading from Malaysia, through the islands of Indonesia – was necessary. Even more remarkable was the hopping leading to a human population in the large island of Madagascar – not from proximate Africa, but rather from Borneo. This remarkable migration[8] was formally over 4,000 miles of ocean but much more probably involved land hopping *en route*.

Once humans had migrated into different niches, inevitably in the longer term they varied both in form and skin pigmentation. For when solar radiation penetrates too deeply, it is life-threatening. So, given mutations within humans, variants with more skin pigmentation were selected in environments closer to the Equator. The characteristics that allow visual differentiation between, say, an Australian Aborigine, a native of Japan, a Swede and a West African were selected. Despite these differences, the commonality of genes among modern humans is very high – estimates of 99.9% are frequently cited. Small differences in our genetic constitutions are sufficient to give rise to differences in our forms, perhaps related to the fact that variations in numbers of copies of quite extended DNA sequences in humans can be remarkably large.[10]

Our past environments moulded not only our forms but also the nature of our survival skills. Diamond emphasises three features – "Guns, Germs and Steel", plus agriculture, that were important in recent selection – in determining which races exercised power over the last thousand years or so.[8] Environments in which metal tools – eventually guns – replaced stone, and in which agriculture involved the domestication of animals and crops, led to societies that gradually dominated the rest. The large human brain became stuffed with more information as society and technology became more complex. Agriculture evolved relatively rapidly in the so-called "Fertile Crescent" – extending from present day south–eastern Turkey to Israel, Syria and contiguous Iraq. Much spawned from there to the Indus valley in the east, and to Europe in the west.

The higher population densities possible within communities with domesticated animals allowed pathogens to pass from them to human hosts – hosts who gradually and necessarily evolved resistance to the associated diseases. Native Americans lacked not only horses and metal weapons, but also resistance to European diseases. It is estimated that diseases spread by Europeans in the Americas killed around 95% of the native pre-Columbian population – which may have been as high as twenty million.[8] The fluctuations that had occurred upon exodus from Africa, resulting in migration to the north or to the east, meant that the distant relatives would meet up again only after a further 70,000 years or so – when Columbus, Cortes and their like arrived in the Americas. Lacking "guns, germs and steel" and horses, natives of the Americas initially regarded the European invaders as gods; the Europeans saw a "savage", to be converted, dominated, exploited or destroyed. Ignorance ensured that neither party recognised not-so-distant relatives.

The selection pressures that led to the relatively recent partial or total demise of some groups of humans[9] are relevant to thoughts regarding the survival probability of societies today – both on global and local scales. Consider the case of tiny Easter Island – approximately 1,300 miles from the nearest land, the Pitcairn Islands. Yet humans arrived there about a thousand or so years ago. However why did people arrive at an area of around sixty square miles (north to south, nine miles wide) when a random voyage only roughly towards the east – in a boat around ten feet in length – would have cast them into a vast ocean of around one million square miles? Were there factors that increased the chance of finding the tiny island from a tiny random probability?

Since the voyage took of the order of two to three weeks, "going back" was unlikely or, at best, extremely hazardous. But the Polynesians could possibly "learn" through many generations of genetic selection, in a manner similar to some birds that evolved to be able to navigate on the scale of global distance. Bird navigation remains a phenomenon only partially understood, but includes guidance from a magnetic compass in their brains. It is automated behaviour, more generally known as intuition.

They could certainly also learn in the classical sense – during their lifetime *via* culture. Tiny islands were detected in this apparently otherwise monstrous ocean uniformity through the sighting of birds, changes in ocean wave patterns and distant clouds. These subtle signs of small and distant land increased the probability of finding Easter Island from around 1 in 10,000 to perhaps 1 in 100. So, given hundreds of years of explorations, the eventual arrival of humans at this speck of land is not so incredible as at first sight; perhaps even probable.

Estimates of the maximum population of Easter Island are, within a factor of two, about 14,000 people. Yet, when Europeans arrived there in 1872, there were only about 100 left. The collapse of this society appears to be primarily due to the over-exploitation of their island resources which, given their isolation, was critical. It may also have been to some degree due to their cultural fixation with the carving of massive statues, for memes leading to self-destruction occasionally blossom – a point nicely encapsulated:

> *Particular features of culture have sometimes merged that reduce Darwinian fitness, at least for a time. Culture can indeed run wild for a while, and even destroy the individuals who foster it.*
>
> – E O Wilson, Consilience – the Unity of Knowledge, 1998

Incursions made by Vikings into Greenland and the far north-east of America proved even more disastrous.[9] These pre-dated those of Columbus by several hundred years, but post-dated the successful migration of humans from Asia to the same continent by more than 10,000 years. So why did the Viking society in the north-east of the Americas and in Greenland collapse? Climate change was part of the problem. From the time of the first settlements (900-1000 AD) until about 1300 AD, Greenland was slightly warmer than now and it was possible to raise some crops. But from 1300, it became progressively colder, and the colony appears to have died out by 1350 AD. Other adverse factors included unsustainable exploitation of the environment, and a cultural bias towards inappropriate European habits. In particular, there was a failure to adopt Inuit practices that clearly worked.

In the collapses with which history is littered, the variables important in Easter Island and in Greenland are prominent, but a total of five that are still relevant today surface frequently.[9] First, a society with a dense population may adversely affect its environment – gross depletion of its forestry, soil and other natural resources with the release of toxins – and hence its food supply. Second, the climate may change. Third, hostile neighbours may emerge, and friendly trade partners lost. Fourth, some groups have more powerful technologies and weapons, and are able to out-compete others. Fifth, where the society might have responded to the causes of adverse change, they have failed to do so. There are clear lessons for forward planning by humanity. Most clearly, the current global population of humans (6.8 billion) is probably approaching an unsustainable level. It would seem that the provision of effective contraception on a global scale should be a top priority. But there is no evidence that this measure is high on the global agenda. And politics muddy the waters of this, and almost every other, suggestion.

The near demise of the Easter Island settlement seemingly provides a nasty portent for an isolated globe with an expanding population that depletes its resources at an increasing rate. But a globe with advanced technology is a far cry from the Easter Island economy. Fusion, which could provide a limitless energy source, is frequently touted as a potential life raft for the species. But there is an enormous technical challenge in constraining the extremely hot plasma that is necessary to achieve fusion. At least initially, it will be simpler to utilise this same process as it already occurs at a safe distance of 93 million miles – to harness solar energy more efficiently.

The evolution of species, and of human variation and recent selection, is an astonishing story. But perhaps even more astonishing is the way, considered next, that molecules built all these bodies – including that of a species so sophisticated that it could travel through space to its neighbouring Moon.

7

Making Bodies from Bits

Cell Specialisation

Teamwork divides the task and doubles the success.

– Traditional saying

The first step in the evolution of multi-cellular organisms was probably the association of single-cell organisms to form colonies in which all the cells performed the same function. Even here, there is a potential advantage for survival – as in shoals of fish, safer inside than at the surface. Once clusters of cells had evolved, some cells within the colony then mutated and, given selection, carried out special functions. In some of these simple cellular clusters, specialization has become so great that, if the organism is disrupted, the specialised cells cannot live independently – they have become committed to teamwork.

The multi-cellular organisms of today come about by successive divisions, along the series 1 –> 2 –> 4 –> 8 –> 16 –> 32 –>… But, if the cells have become specialised in their functions, this begs a question: how can a common inheritance allow the expression of these differences? Evolution found a simple solution. Every cell type receives the whole complement of genes required to produce the complete organism. But, for cells with specialised function, only those genes necessary for that function are expressed. In sexual reproduction, the starting point of the single cell is generated by the fusion of information from female and male. In humans and our mammalian relatives, selective switches then act during development to turn them into cells of thyroid, bone, skin, etc.[1,2] Over a few billion years, specialised cells engineered the solutions required for sophisticated machines.

That the simple solutions of engineering find their place in biological machines was beautifully illustrated by D'Arcy Thompson in his book *On Growth and Form*.[3] Molecules in bodies, and organisms in their societies, have found strikingly similar solutions in the search for function.

Ropes evolved in human societies to connect objects together, and are given greater strength by wrapping numerous fibres together in a helix. Collagen is Nature's rope, providing the connective issue within animals, and particularly within mammals where it is the most abundant protein – a major constituent of tendons, cartilage and skin. The forms of flexible shower pipes, and of the collagen helices found in the tail of the rat, are amazingly alike. Like the ropes of society, collagen is composed of long strings (polymers), three of which are wrapped round each other in a helix. It is quite precisely the rope of our internal structure.

Evolution had to not only produce "one-dimensional" ropes but also "two-dimensional" walls. This required not simply collecting small molecules together to give long strings, but also the cross-linking of these strings into nets. This is precisely how the wall of a bacterial cell is formed. Small sugar molecules are connected in long strings in one dimension, and these strings are then cross-linked by amino acids in a second. Nature discovered warp and weft long before the evolution of cloth. Bending the molecular net into a third dimension allowed the membrane of the bacterial cell to be surrounded by a container – a cell wall – resembling the surface of a football.

Once food molecules arrive at the surface of a cell, the simplest solution to the problem of getting them inside the cell is simply to invaginate them. A more sophisticated solution is to "pump" them across the membrane through specialised channels constituted from protein. The reverse process is used to expel the waste products from the cell. The channel-containing membranes are like the walls of a medieval city. Both constructs protect the internal contents from external dangers; both close and open ports to let in essential supplies or to remove toxic waste; both can accommodate sensors to detect attacking entities, so that a defence system can be activated.

In a factory, the admission of supplies into one area, and export of products from the same region, would be messy. In the evolution of organisms, the molecular structures of biology encountered the same problem. The solution was the evolution of mouth and anus, moulding the form of animals a few hundred million years ago. Further simple engineering solutions were required if multi-cellular organisms resembling worms, and their more complex successors, were to evolve. In organisms containing gargantuan numbers of cells, the length of the path from cells that form the lining of the digestive tract to the surface "skin" is necessarily large compared to the dimensions of cells. So circulation systems, with veins and arteries, evolved to allow deeply buried cells to get their nutrients and to remove their waste. In human societies, roads are the structures analogous; trucks, with their attendant loads, analogous to the molecules that flow along them.

But transport along a vein is a relatively slow process. Yet withdrawal of cells from life-threatening heat and other dangers must be extremely rapid – satisfied by the evolution of rapid electrical pulses along nerve cells. Electricity cables analogously transmit rapid signals in human societies, but the evolution of these cables was inevitably much later than those of nerves.

Long organisms could evolve through the replication of a number of identical, or similar, building blocks – segments. In the initial stage of such developments, the requirements of genetic change are minimal – simply duplicate. This turned out to be such a successful solution in the building of long bodies that it is found in worms, insects and vertebrates. And why not, for this is another engineering solution found in human society – in the construction of passenger trains. In the human spine, very similar vertebrae are constructed many times in a manner that facilitates the production of a body that is longer than it is wide. Research with fruit flies provides dramatic evidence for the method of production of similar body segments – all under the control of a simple genetic switch. The body plan found in fruit flies – and more readily seen in bees – is common throughout the insect kingdom.

The analogies are everywhere: brains/computers, eyes/cameras, voice box/loudspeakers, lungs/trees, liver/fermentation chamber, and many others. The forms and devices required for the efficient functioning of machines are "out there" – given the forces, and the environments, which operate at the surface of the Earth, there are optimal solutions. It is simply a matter of the time taken by evolution to find them. Molecules inevitably found them first, and made them faster, smaller and more sophisticated than the devices of human societies. Humans, although late arrivals at the ball, will in the long run happen upon their best solutions – once more through variation, fitness and selection.

Generating Structural Variety

Worms, insects and vertebrates may all use the repetition of very similar constructs in the construction of long bodies. But the last two are much more heterogeneous structures than are worms. How was this possible?

If flies are exposed to X-rays, mutations occur in their DNA. This is a bit like spraying rubber bullets into a crowd of rioters – it is not known beforehand whether serious damage will occur; nor, if serious damage does occur, at what site it will be found. If irradiation is carried out early in the life cycle of the fly then, on occasions, the result is an amazing disturbance in the body plan of the adult fly that subsequently develops. For example, one mutant has legs growing out from the point at which antennae are normally attached. It is as though, in the assembly of a jet plane, the instruction "site a radar unit in the cockpit" had been replaced by "site a set of wheels in the cockpit". In another mutant, two pairs of wings, rather than the normal single pair, are produced. These effects show that at some stages in the development of complex organisms, relatively simple principles are involved.[4]

A large body of research leads to the following conclusions. The fly embryo contains fourteen segments. If such segments all expressed only very similar genes then they could be used to construct something like the central portion of a worm, a spine or repeated segments of an insect thorax. However, of the fourteen segments, three lead to the formation of head segments, three to the formation of thoracic ("middle") segments, and eight to the formation of abdominal segments. Under the pressure of evolutionary selection, very different genes have come to be expressed in these three structural components – even though they derive from fourteen repeated segments in the early embryo.

During development, cells are allocated to the segments of the embryo. However, they are also instructed, through the expression of only a few genes (control genes), as to whether they will produce antennae, legs, thorax or abdomen. So, although all the cells have a complete set of plans, some are instructed to express only antennae plans, and some only leg plans, thorax plans or abdomen plans. Amazingly, the control genes are the *same* in flies as humans. The control, or "powerful manager", genes evolved prior to the existence of the organism that was the last common ancestor of humans and flies. That common ancestor lived about 600 million years ago – a shout for Darwin if ever there was one.

The above conclusions are reinforced by observations on diverse animals. The early stages of development give rise to *remarkably* similar embryos in species as diverse as fishes, amphibians, birds and mammals. These observations suggest strongly that early stages in embryo development are based on patterns that were established early in evolutionary history. The human embryo is in fact segmented like the fruit fly larva, according to a basic plan that is valid for the building of all complex bodies. More evidence that all life derives from the same spring: when you consume a lettuce, or kill a fly, a distant relative meets its demise.

Why are there very ancient and common instructions for making these differing parts of bodies that are formed early in development? Once an early and fundamental instruction is in place, it becomes not only finely honed but also difficult to displace without great risk to development. If a non-neutral mutation occurs in such "body plan" genes then what follows is likely to have a strongly adverse effect in the battle for survival. Thus, there should be strong discrimination against changes to the early-evolved pattern. Analogy with the automobile industry again offers insights. Fundamental parts (wheels, engine, steering wheel, axles, gearbox, clutch) were adopted early in the evolution of the vehicle and are still retained today. But an enormous amount of sophisticated detail (eg multiple electrical circuits, computer control) has evolved since that time without changing the basic plan. In the case of molecular biology, it is this sophisticated detail that makes flies, mice and humans so different. But the early starting points are based on the same mechanisms.

Putting the above arguments in another way, the option of starting again is not on the cards. Newly evolved forms of life are very likely to be less successful in function than are the forms that have evolved over a longer period. Yet again, analogies are found in human engineering. In the latter half of the twentieth century, the Wankel engine was touted as a replacement for the conventional combustion engine, but it did not take over. Whatever the relative merits of the two forms of engines, the Wankel design suffered from a major problem: that of displacing the established way of doing things. Tooling up for a completely new start is a costly and hazardous business. Displacing old mindsets and systems can be difficult, which is why ancient institutions may suffer from being saddled with outmoded statutes, regulations and long-established traditions. But Nature also suggests that there are often advantages in retaining ancient systems.

Evolutionary biology is even more constrained than is the evolution of human engineering and institutions. For, in biology, there are myriad layers of complexity that were laid down hundreds of millions of years ago. Organisms are built as might be dictated by a large committee operating under archaic rules.

Embryonic Development

It is not birth, marriage or death, but gastrulation which is truly the most important time in your life.

– Louis Wolpert

The development of a human posed mind-boggling problems for cultures over the centuries. One theory got round the problem by proposing that the embryo was already present in the egg, and then simply grew at the appropriate time. Thus, the embryos for all past and further generations had to be pre-formed, one inside the other. This theory stretched human imagination on the point of how many generations could be pre-packaged inside each other in this way. But nothing better was available at the time – a model based not on evidence but on ignorance. Better proposals could be made only with new knowledge. Much is now known,[1,2] but great ignorance still remains.

Some cells of a female are selected for development into egg cells even when she herself is at the early stage of development as an embryo. Molecular signals that lie dormant until a little more than a decade after birth are required for egg maturation; signals that lie dormant for a similar period, but then also transmit their messages monthly, are required for ovulation. The development of a baby, consisting of tens of trillions of cells, from the single cell of a fertilised egg in nine months emphasises further how Nature and its clocks humble our efforts and understanding.

Fertilised eggs appear to have close to spherical symmetry, but the embryo into which they develop is more asymmetric. How is this achieved? The fertilised egg is, in the mother's body, in an asymmetric environment – different portions of its surface have different environments. One portion of the egg interacts with a protein produced by a maternal gene, and one end of the egg thereby becomes different to the other – as for the Earth, with its different poles. The embryo can therefore be instructed to have parts that will end up as the head or the tail. In a similar manner, information is received from the mother that leads to a front and a back in the developing embryo. There is also evidence for at least one other cue for asymmetry: that of the direction of an electric field. This cue is internal to the egg system (rather than coming from the mother) and is present when an embryo is at the stage of only four cells.

In the formation of a tadpole, the fertilised egg has divided 6 times into 64 (2^6) cells after 4 hours, and into around 10,000 cells after 6 hours. The first 12 cleavages (to give 2^{12}, or approximately 4000, cells) occur synchronously. It is the further cell divisions – into about 170,000 cells after 32 hours, and into a feeding tadpole (*ca* 1,000,000 cells) after 110 hours – that cause a marked increase in mass. Since the starting point was only one cell type, yet the developed animal contains many cell types, the cell divisions must on occasions be asymmetric. Asymmetric cell division is achieved when, despite DNA being fully copied into both daughter cells, the proteins within the parent cell are not equally partitioned.

The fertilised single human egg cell divides in a similar manner, but more slowly. After 6 days, it has divided into somewhere in the range of 32 to 128 undifferentiated cells. About day 10, the cellular mass embeds itself into the wall of the womb and becomes surrounded by membranes on the outside. These membranes carry food and oxygen to the cells. The actual area of the cellular mass at day 10 is about one-tenth of that of the surface of a pinhead. The embryo is about 0.5cm long at the end of week 4 and is beginning to show characteristic features. After only 5 to 6 weeks, the faint pulsing of a nascent heart can be detected.

After only two or three cell divisions, each of the resulting cells is remarkably adaptable, insofar as each one retains the potential to develop into a normal adult. Thus, if a mouse embryo is taken at the two-cell stage, and one of these cells destroyed, the remaining cell will frequently develop into a normal adult. An even more remarkable result is obtained if two eight-cell embryos are combined. If one of these embryos comes from a strain with black fur, and the other from a strain with white fur, then a piebald black and white mouse is produced. Such animals – chimeras – have four parents; they are otherwise normal.

At the early stages of cell division, the structure – a blastula – is best described as a fluid-filled ball of cells. There next occurs, in the development of all animals, a dramatic change – gastrulation. Some of the cells migrate up the *inner* lining of the ball. To do this, they pull themselves along using long, thin appendages with sticky tips. They eventually form connective tissue, heart, blood cells, bone and smooth muscles – all cell types that, in the absence of injury, have little contact with the external world.

Other cells, from the *external* part of the embryo, not only eventually provide skin but also move to form the anus to mouth connection. These latter cells of the nascent gut also eventually provide salivary glands, liver, lungs and the nervous system. They have in common more extensive contact with things and influences from the external environment.

What might be inferred from these observations? The similar forms of early foetuses of very different species suggest that, in these animals, cells migrate and differentiate according to ancient and common programmes. Gastrulation, being a very early stage of animal development, should thereby be a very ancient process, showing – as it does – common features among widely different species. Thus, the patterns of cell migrations should reflect the functions that these cells performed early in the evolution of life – providing internal services *vs* handling things from the environment – as seems to be the case.

In support of the above picture, a structure that pre-dates another structure in evolutionary terms typically appears earlier in the embryo. For example, the backbone, the common structure among all vertebrates such as fish, reptiles and mammals, appears as one of the earliest structures laid out in all vertebrate embryos. Conversely, late-evolved structures appear late during embryonic development; as in humans, where the most sophisticated part of the brain is late to develop. Additionally, if an ancient structure has later become redundant in a more recently evolved species then it appears in the early embryo, but later recedes – as happens when humans develop a coccyx rather than a tail.

All this makes good evolutionary sense – recent changes in form and function affect late stages in a sequentially expressed program that has developed over a time of the order of a billion years. The germs of this idea stemmed from theories of Johann Friedrich Meckel (1781-1833), Etienne Serres (1786-1868) and Ernst Haekel (1834-1919). Although Haekel rather over-egged his theories, the cell migrations that occur in gastrulation do appear to inform us about what happened in a very primitive ancestor. Embryonic development contains useful information regarding the tree of life.[5]

In these divisions, movements and specialisations of cells, a continuum with the laws that govern the behaviour of inanimate matter may appear to have been broken. But it has not for, in these processes, the forces and principles described in the preceding chapters provide the driving processes. These forces and principles do not simply apply to physics and chemistry. Given a continuing supply of energy, they naturally lead to the laws of biology, to new sets of interactions that give rise to new aggregates of matter, and to increased complexity.

In the production of the head, thorax and abdomen in insects, it was seen that segments evolved to produce different structures according to the different sets of genes that are switched on within them. But how can different sets of genes be switched on according to the *position* of cells in the developing embryo?

Suppose that six cells lying along the length of an embryo are required to produce six different structures: A, B…or F. These A, B… or F structures are produced by initiation of the expression of cascades of different genes. A "master gene" initiates the expression of these different cascades. There are very few of these master genes, and they have common origins throughout the animal kingdom. They are able to trigger very different events by acting according to a *concentration gradient* of their protein product – a product that naturally gets more diluted at greater distances from its origin. In practice, *large groups* of cells can be instructed to make products of type A, B…or F according to their position along the gradient of the diffusing substance. But the simple model illustrates the principle.

How bodies survive, how they are sculpted, what they can do, and why they behave as they do, all so relevant to our state and decision-making, is considered in the following chapters.

8

The Needs and Forms of Bodies

Sensing the World

A beautiful thing never gives so much pain as does failing to hear and see it.

– Michelangelo

Plants are relatively immobile and vulnerable to predation by browsers. They are therefore adapted to grow replacement parts when a portion is eaten. The perception of pain is therefore unnecessary in plants – they do not require a nervous system, nor the eyes, ears and muscles that allow rapid responses to external signals. But some plants have limited, and specialised, mobility. The sunflower orients itself to benefit optimally from the Sun's radiation, and the sundew (Venus's flytrap) closes its "jaws" to capture insects. Both reactions are initiated by primitive senses.

In contrast, the lifestyle into which browsers and carnivores evolved demanded rapid reaction and movement. Their sophisticated senses evolved from humble beginnings – evident in bacteria, in which receptor proteins bind molecules that occur in their environment and so induce movement within the proteins. These changes allow overall motion of the bacteria – towards food particles, away from toxins – and are prototypes for the evolution of taste and smell. And the prototypes have evolved into awesome systems, for a mouse makes over 1,000 receptor proteins in its olfactory system.

For the remaining three traditional human senses, photons are detected for sight, sound waves for hearing, and a force in the sense of touch. The Nobel laureate Martin Chalfie has shown that, even in a humble worm, the sense of touch involves the protein products from at least twelve genes – a number certain to be dwarfed in humans with their exquisite sensitivity to materials of subtly differing textures. And if the organism is to derive optimised benefit from its senses, they must convey a realistic and self-consistent impression of the environment. And they normally do, for a hexagonal doorknob can be concluded to be such either by touch when blindfolded, or by sight at a distance.

Appendages

After that he kissed each of the fairies, and ran home as fast as his little legs would carry him, singing all the way, 'Humpty Dumpty! Humpty Dumpty!'...

– 'Old Hungarian Fairy Tales' by Baroness Orczy

The automated sensing devices of Nature are wondrous: a kiss tender, rather than forcing; and, if performed standing on two small feet, benefiting from an amazing sense of balance – even more amazing when maintained in running.

Animals commonly derive advantage from rapid and variable movement. Mobile appendages evolved about 450 million years ago, although their prototype may have arrived much earlier – "pods" whose function was primarily that of sensing the world. In many animals, four and six appendages have appeared as common effective solutions for standing on land – as also in chairs and tables – and also for rapid movement. Standing on two appendages – birds, kangaroos and humans – is more demanding, facilitated by an orthogonal twist and spread of their ends, plus a finely honed sense of balance. Shaping complex appendages is an interesting business.

Sculpting for New Needs

I saw the angel in the marble and I carved until I set him free.

– Michelangelo

I choose a block of marble and chop off whatever I don't need.

– Auguste Rodin

In configuring detail, Nature often works like Michelangelo and Rodin. She sculpts.

Microscopic examination of human cells shows that we are constituted from somewhat more than 200 markedly different types (although modern molecular biology shows that there is, in addition, a large number of cellular sub-types). In each of these 200 cell types, a different part of the DNA program is switched on. Parts of the DNA program also tell a cell its appropriate position in the organism, and additionally the shape and size of the structure of which it is part. How cells are instructed with regard to their position in an organism has come from experiments carried out on the development of limbs.[1]

In birds, the front limbs evolved to bear the wings. The wing of the chicken contains three rather differing "fingers" (digits, labelled here 1, 2, 3). The early embryonic structure that develops into the wing is known as a wing bud. It has a region of "instructing cells" at one side of the bud. These instruct the digits to form in the order 1, 2, 3, where 1 is nearest to the instructing cells. This is known from an experiment in which these instructing cells were taken from one bud and then added to the *opposite* edge of another wing bud. Thus one wing bud had been given *two* sets of instructing cells – one at each of its edges. The result was the formation of *two* sets of digits on the one bud: one group of three appearing as the mirror image of the other group of three (passing across the bud, what grows is 1, 2, 3, 3, 2, 1). Additionally, the master signals from the position-instructing cells determine in a coherent manner the pattern found within each of the digital structures – the appropriate expression of cells that make up cartilage, feathers, muscles, nerves and tendons.

So, something produced by the instructing cells must tell the remaining cells of the wing bud what genes to express. That "something" is retinoic acid which diffuses out from the instructing cells to cause the formation of the appropriate digits at the appropriate place. A high concentration of the diffusing molecule dictates digit 1, and decreasing concentrations at greater distances from the producing cells successively dictate digits 2 and 3. The cells in the centre of the bud wait for the instructing cells to tell them what to become.

The above experiments support a general applicability of the earlier introduced model of concentration gradients to instruct the development of complex bodies. Analogous instructing cells may be taken from the corresponding regions of human, or turtle, limb buds. When these are transferred into a chicken limb bud, they produce the same effect as the chicken's own cells. The instructing mechanism is the same, or almost the same, in humans, turtles and chicks. Therefore, it is very likely to be ancient.

Two methods could, in principle, be followed in the production of digits from the bud. Either (i) build each digit from the bud as a discrete entity; or (ii) build the whole bud forward and then "sculpt out" the unwanted cells between the incipient digits – either build as does a bricklayer, or as does a sculptor. There are clues as to how Nature will do it, for she operates under archaic rules (chapter 7). Making a house from bricks is a relatively modern technique, developed with prior knowledge of the point to where you would like to get. Primitive man – like Nature – did not have knowledge of "what was coming next", but could enhance, or even create from scratch, a cave through "sculpting out". Our guess is that Nature will sculpt – and she does.

In digit construction, the sculpting out is achieved through *apoptosis* – programmed cell death. Featureless buds evolved into sophisticated hands through sculpture with cells. Nature builds a featureless mound and then wipes out the cells that are not needed. She does indeed work in the same way as Michelangelo and Auguste Rodin. And the extent to which she does became even clearer from the pioneering work of Sydney Brenner and John Sulston. It was John Sulston who, peering down a microscope, observed the precision of the sculpting. He told me that, on visiting Darwin's house, it moved him to see the worn carpet below the great man's chair – a product of countless hours of dedicated labour going on above, which reminded him of similar wear below his own microscope chair.

Sydney Brenner had seen the need for an attack on development at a fundamental level:

> *There are two great problems left to science. One is development and the other is consciousness. The reason that neither has been solved is that the people working on them are stupid. But now we have cracked the genetic code, we are going to finish them. Francis [Crick] is going to take consciousness and I am going to take development. We will have them solved in ten years. Who wants to come along for the ride?*
>
> – Sydney Brenner, 1963

The above quote, made to students in a pitch to persuade the brightest of them to become molecular biologists, reflects Sydney in expansive and provocative mood. Despite the deliberate overstatement, he was – as often – putting his finger on the nub of the matter. Francis Crick did indeed make notable contributions to our understanding of consciousness in the following decades. But that problem is the more difficult one, and still remains very poorly understood. Sydney Brenner also formulated his views in somewhat less trenchant terms:

> *I remember that we decided against working on animal viruses, on the structure of ribosomes, on membranes and other similar trivial problems in molecular biology. I had come to believe that*

*most of molecular biology had become inevitable and that, as
I put it in a draft paper, 'we must move on to other problems
of biology which are new, mysterious and exciting.' Broadly
speaking, the fields which we should now enter are development
and the nervous system.*

In a human adult, more than a thousand billion cells are made every day, and
a similar number "commit suicide" through the process of programmed
cell death. In the work of Brenner and Sulston, key discoveries about
programmed cell death were made by using a worm about 1mm long –
the nematode *Caenorhabditis elegans*. It was the idea of Sydney Brenner to
follow development in the worm since mammals are too complex to follow
the details. The worm is conveniently transparent, allowing development
to be followed under the microscope. And what astonishes is that about
35% of its genes have human analogues.

Detailed studies on the worm, providing understanding as to how
also *we* are made, showed that, out of a total of 1090 cells, precisely 131
reproducibly die during development.[2] Each and every worm undergoes
exactly the same program of cell division and differentiation, and the
adult ends up with 959 cells. Moreover, this programmed cell death is
controlled by a unique set of genes, and the corresponding genes exist
in more complex organisms, including humans – as shown by Robert
Horvitz in Cambridge, Massachusetts. Cell death is a required aspect of
normal development, and sculpting is common throughout the animal
kingdoms. Nature builds a featureless brick wall and then knocks out
some bricks to make doors and windows. For their work, Brenner,
Sulston and Horvitz shared the 2002 Nobel Prize in Physiology or
Medicine.

Changing Faces

Mutations within genes not only produce proteins that are new and
express their functions directly. They also lead to new proteins that
control the *times* at which large sets of genes produce their characteristic
proteins. The pressures for selection for such times of expression may be
great. In the presence of carnivores in the environment of the open plains,
the infant gnu must get moving quickly – otherwise a brutal selection
will ensure it is not with us for long. As a consequence, the genes that
instruct for "standing up" in the infant gnu are expressed immediately
following birth. Within six million years, the collective genes that code

for "standing up" in the newborn of *Homo sapiens* have become delayed in their expression. Through selection, different lifestyles result in different genetic modifications.

Neoteny is the retention of juvenile characteristics in the adult form of an animal. There is a delay in the development of some body parts and functions, which arises from the delayed expression of large sets of genes. There may be a striking example of neoteny in the development of humans. In the approximately six million years since we shared a common ancestor with chimpanzees, "manager genes" are among those that have been modified, and so in today's species there are differences in the degree of expression of whole groups of genes.[3,4] Hence, small overall differences in DNA sequences can result in relatively large differences between the species – in a similar way to which a new manager can change the output of a factory by selectively influencing the lines of production.

In facial appearance, a baby chimpanzee is much closer to humans than is the adult (Figure 8-1).[5,6] But a quite different adult face is produced in the two species without large changes in their totality of DNA sequences, suggesting changes in a few genes that act as managers. Apparently, in our nearest relative, the genes that code for the raised brow bone, and for the protrusion of the jaw and nose, are expressed to a greater degree in the adult chimpanzee than in adult humans. Although it is fanciful to describe *Homo sapiens* as an ape that never grew up,[5,6] new faces are generated by small changes within the factory.

Figure 8-1. Baby and adult chimpanzee (from Naef, 1926)[5]

Over the past few years there has been panic in some quarters because scientists are looking to use very small particles, and to build very small machines – the field of nano-science. The building of micro-devices – particularly those based upon miniaturisation in electronics – represents a spectacular advance. However, some have proposed that "nano-robots" will be manufactured from tiny components; robots able to roam the body, carrying out repair jobs. Currently, this is the realm of science fiction; likely to remain so for a long time. Because a few billion years of evolution have already produced much smaller things – the molecules of biology – that carry out these jobs with astounding efficiency.

Biology strides like a colossus above human technology. Some bacteria swim by rotating helical filaments – to be likened to tiny legs – distributed on their surface. The rotation of each filament is driven at its base by a rotary motor that is only around one hundred thousandth of a centimetre in diameter, and is made from about twenty different parts. When the bacteria detect a concentration gradient of food in their environment, the direction of rotation of the motors is controlled so that the bacteria can swim towards the higher concentration. The "micro-systems" of humanity are pathetic monsters in comparison to these devices. Organisms exploited the ultimate nano-science billions of years ago, and the functioning of all organisms relies on it.

The minute size of the vast majority of cells is possible because the engineering of organisms commenced with tiny assemblies of molecules. The small diameter of a cell requires, relative to a larger construct, a large surface area to volume ratio – illustrated by sub-division of a one-metre cube of wood. Before division, it has a surface area of six square metres. Upon division into 1mm cubes, a thousand million such cubes are produced, with a total surface area of 6,000 square metres – yet with an unchanged total volume.

If the wood were placed in water then it could be saturated with water enormously more quickly in the divided than the undivided state – reflected in the performance of human red blood cells, as pointed out to me by Philip Kuchel of the University of Sydney. The red blood cells carry oxygen from the lungs to other body cells, and also carry carbon dioxide from these body cells back to the lungs for exhalation. Almost half of the volume of our blood is constituted from red cells, and thus we have about 2.5 kg of red cells in about 5 litres of blood. The cells are so tiny – roughly 0.005 mm in diameter – that the mass of 2.5 kg is made up from around 5 thousand billion of them. An average red cell survives for roughly 120 days in the body, the "worn out" ones being destroyed by other kinds of cells. As a consequence, we need to manufacture almost half a million new red cells *each second* to maintain the population.

Knowing the rate at which blood is pumped round the body, and the average lifetime of the red cell, it is possible to calculate that a typical red cell travels 300 kilometres during its lifetime. Even more strikingly, it can be shown that the water content of a red cell is exchanged (with water from outside the red cell membrane) at least 50 times every second. This phenomenal activity is possible because, the cell being so small, its surface area is large relative to its volume. The rapid exchange of water means that our 2.5 kg of red cells absorb (and excrete, thereby maintaining the same cell volume) *around 8 million litres of water per day.* This is the amount of water required to fill about 3 Olympic-size swimming pools. We must already have a sense of awe. But our journey still has far to go.

Communication within Large Bodies

We are confident it will be done quickly and efficiently in keeping with the high standards of accomplishment set by the Merchant Marine.

– Fleet Admiral Chester W Nimitz, US Navy, Chief of Naval Operations in WWII, on the transport of war materials over large distances

The transfers required by Admiral Nimitz were of relatively large masses over large distances and took times of the order of days and weeks. Communications and transfers within giant aggregates of cells involve much smaller masses and distances, and can accordingly be achieved more quickly. How fast are signals transmitted within mammals?

The cell in which an environmental effect is sensed – say, the foot – may be at a considerable distance from where it is processed – the head. Therefore, conversely, beneficial actions of the foot require signals to be transmitted over large distances from the head. When a limb comes into contact with a red-hot surface, or an impala reacts to avoid the jaws of a lion, a rapid transfer of information is necessary. This requirement led to the evolution of long, thin nerve cells – up to about a metre long in humans, and several metres in the giraffe – that work through the transmission of a pulse of electricity.

The electrical pulses travel quickly – as they need to. In playing tennis, the biological machine *Homo sapiens* senses a ball travelling at speeds in excess of 45 metres per second (100 mph). It then carries out an automated and complex internal processing of signals, and responds with such precision so as to be able to return the ball with a positional accuracy of about 10cm at a distance of around 10 metres. The position of the racquet is not only adjusted by arm movement, but also by the coordinated movement of the legs, while the fourth limb can be used as a balancing device. The response time for locating the position of the ball, estimating the velocity of the ball, and reacting *via* signal from eye to brain, and then from brain to many different and distant sets of muscles is of the order of a few hundred milliseconds.

In other cases, the needs of the organism can be satisfied by transmission of the information on a longer timescale – several seconds, or longer. Such a need can be achieved through the circulation of a liquid (eg the blood) that contains the signalling molecule. All these responses – fast or slow – are achieved by a machine built from an egg that has a diameter of about one hundredth of a centimetre, and a sperm that has a width about twenty times smaller. The analysis may be turgid but engenders awe. Some matter has followed a remarkable journey from the chaos of the Big Bang.

Pigs Don't Fly

In 1991, the distinguished paleontologist and evolutionist Stephen J Gould, famous for his elegant prose, published a book *Wonderful Life*.[7] In it, he made the case that if the evolutionary experiment ("the tape of life") could be run again from its early stages then the outcome would probably be quite different. The case seemed convincing.

Gould's stance is illustrated by reference to the enormous diversity of life found in the fossils of the Burgess shale, located in British Columbia and providing the most detailed evidence of the "Cambrian explosion" which occurred about 500 million years ago. He considered the importance of one-off coincidences – "contingencies" – that occur during the course of evolution. Should a meteor have impacted upon the immediate precursor of the enormous variety of vertebrates now occupying the Earth, the precursors of *Homo sapiens* and all other vertebrate animals would have been wiped out. Gould's conclusion was that humanity would not have come about. Unsurprisingly, his view was given much media coverage at the time. Gould was famous, wrote well and knew much about the fossil record. He was also an accessible "media man". So, he's probably right.

However, on occasions, others then come along and show that what has been proposed may in fact be bunkum. In his book *Life's*

Solution,[8] Simon Conway Morris, another distinguished paleontologist, convincingly rebutted Gould's arguments. Conway Morris argued that if the first creatures with bones were wiped out, bones would arise again simply because they are too good to miss out on. His thesis is supported by the observations of D'Arcy Thompson (chapter 7) that the solutions of engineers are commonly found in animals. This being so, evolution should show repeated signs of convergence upon "good solutions".

Simon Conway Morris's conclusions are based on much evidence indicating that evolution is often *remarkably* convergent[8] – demonstrable at the levels of (i) the molecules of biology, (ii) the organs of organisms, (iii) the forms of organisms, and (iv) cultures of organisms.

(i) Molecules: Robots for finalising details of car assembly, if constructed by independent companies, may differ in the routes taken by the robotic arms from the base of the robot to the car. But cars are so close to a standard shape that the interactions between the robotic arms and the largely constructed cars must occur at very similar points in space – at these points, there must be convergence in the various designs. This restriction bears analogy to the robotic assembly of enzymes. An enzyme is a molecular robot that must recognise and act upon a specific shape (eg alcohol) to metabolise it. The trial-and-error approach of evolution sometimes led to differing strings of amino acids that meandered through different points in space (cf different shapes of a robot arm), but ended up touching the molecule to be metabolised (cf the car) in the same way.

Thus, although the *guiding strings* of the amino acids in the enzyme have different structures and navigate through different points in space, they orient the key amino acids involved in catalysis to *the same points in space*. In some cases, bacterial and mammalian enzymes have very different amino acid sequences, but the three amino acids that do the job of catalysis are the same in both enzymes and have similar orientations in space.[9] The required solution was found by completely different routes. This is a remarkable convergence in some molecules that provide organisms with function.

(ii) Organs: The blindfolded dolphin that locates and collects rings thrown into a large pool is a familiar entertainment. Its ability to emit high-frequency clicks, and to use the echoes to find the object, goes hand in hand with an enlarged cerebellum; perhaps no coincidence, since another mammal – the bat – also famous for this ability, also has an enlarged cerebellum. Not only are certain cognitive abilities convergent but these abilities are also associated with convergent features of organs.

(iii) Organisms: The latest common ancestor of the placental and marsupial mammals is estimated to have existed slightly more than 100 million years ago. In size and form, this common ancestor was closer to a shrew than to a large mammal. Nevertheless, a marsupial sabre-toothed tiger evolved as did a placental sabre-toothed tiger. The two are amazingly similar in form;[8] given about 100 million years, evolution produced this form independently in the northern and southern hemispheres. The form of such a predator is evidently a "standard engineering solution".

(iv) Cultures: Human agriculture involves transport, weeding, fertilisers and cropping. As vividly described by Simon Conway Morris,[8] leaf-cutting ants, though very different organisms from Homo sapiens, do all of this. Some ants collect and cut leaves. Others then collect the pieces, cut them into smaller parts and carry them to an "ant road". A third group then transports the pieces to the nest. The leaves are not used directly but instead "mulched" and mixed with a fungus. Enzymes of the fungus break down the cellulose in the leaves, and thus food for the ants is provided. Fertiliser is applied in the form of nitrogen-rich manure.

The ants also weed the fungal gardens by removing spores of an unwanted fungus. But these alien spores are also attacked in a second approach. The leaf-cutting ants use antibiotics – from the same group of microbes as used in human medicine – to kill these alien spores which otherwise ruin their gardens. Ants used antibiotics before humans. And it doesn't end there: bolas spiders displayed their skills long before gauchos exploited a similar technique in South America; anglerfish used their specialised skills long before fishermen.

Convergence abounds within molecular structures, organs, organisms and cultures. But there are occasionally pitfalls regarding their evolution. It has been widely disseminated, and believed, that eyes – clearly a "useful solution" – evolved independently no less than forty times. But, astonishingly, key genes that dictate the positions of the very different eyes of mice and of insects are so similar that they can be exchanged and still function – perhaps a suggestion that the eye evolved only once? There are persuasive arguments[10] that the "once" *vs* "many times" dichotomy is rather silly.

There can be little doubt that if the automobile were to evolve again from scratch, the end product would be similar. But did the automobile evolve once or many times? First, you must try to define an automobile, and that is not easy. Try "a device with four wheels and an engine". So, was Uncle Herbert's trap, with an added engine driven by steam, an automobile? Similar problems arise in attempts to define an eye – at the minimalist level, is it sufficient to require only a system to detect light and for the organism to derive benefit from this detection? Clearly not, for even plants have this ability. But the detection of light, and response to this detection, indicates devices that may be called upon in the eventual evolution of eyes.

In several places in the world, "automobile-like" structures evolved, and their unifying principle was that they consisted of a collection of devices that were similar. In a manner that is analogous, combinations of the products of large numbers of genes could result in "eye-like" organs. Some of those gene products did other things earlier in evolutionary history, just as axles preceded the car. In a nutshell, "it is about the components". Different components have different histories, and this cannot be accounted for in the simple dichotomy of "one *versus* more than one origin".[10]

Convergence is so palpably true that it is easy to go over the top in evoking the principle. There is not convergence upon a few forms of organisms: tiny bacteria on one hand, through beetles, to whales on the other. It was Darwin himself who pointed out the reason for this diversity: there is great advantage in having your own niche, and the exploitation of very different niches ensures diversity of form. But, given the laws of the Universe, the evolution of life, specified niches and sufficient time, then certain kinds of solutions seem inevitable. And certain solutions precluded for, as emphasised by Simon Conway Morris,[8] evolution of a biological jet engine to work in the atmosphere would be difficult in the extreme (even though squids exploit the principle in a less demanding medium – water). So pigs are rather too big to fly.

But whether bacteria, pigs or humans, they must be controlled, serviced and defended; and, recently, their manipulation has reached new levels of intensity. How is considered next.

9

Running and Manipulating
the Human Machine

The Nano-Controllers

Actions, whether simply internally controlled or also requiring environmental prompts, are triggered by molecules.[1] If Fred makes a violent threat to Peter, adrenalin is released into the latter's bloodstream. The adrenalin molecule, with a length of about one thousand millionth of a metre, prepares the body for "fight or flight": the heart beats more quickly, and so blood is pumped more quickly round the body; energy-providing molecules delivered at a faster rate to their site of action in muscles. Additionally, it is costly to store something that is immediately needed when a crisis is at hand. So, a molecule that is normally used to convert blood glucose to a form in which it can be stored is temporarily inactivated. These responses are automatically activated to promote survival; the individual has no choice. The body is a survival machine produced and controlled by molecular engineering. Molecules enable our *every* action. When they bind to each other, an aggregate with new properties is formed, and this new entity can trigger new events.[1]

Supply

Production requires the delivery of raw materials to the site of manufacture. In the case of an animal, these supplies are minerals and organic molecules, and oxygen molecules to "burn" (metabolise) the organic molecules to produce energy. For a human consisting of *ca* 10^{14} cells, many of the cells must be buried deep inside the cellular mass. So the delivery of supplies required the evolution of sophisticated delivery mechanisms. The delivery of oxygen to every one of around 100 thousand billion cells in the body is a case in point. Oxygen is absorbed from the lungs into the blood stream by a molecule – haemoglobin – to which oxygen binds. The blood is then carried all over the body by the vascular system, and the oxygen dropped off at the appropriate points.

What is achieved in this delivery dwarfs delivery systems devised by humans. The mail delivery systems of the world reach only a few billion individuals whereas the circulation of any one of us reaches around 100 thousand times as many cells. And the delivery is much more continuous than of mail, for a brain cell starved of oxygen for ten minutes may end up dead. Over the first four decades of life, a pump with a remarkably low failure rate drives the deliverers. Although clogged arteries – "snowed-up" roads – may hamper, or even eventually foil, their efforts, the molecules do not take "industrial action". They deposit their parcels in letter boxes that are typically less than one hundredth of a millimetre in length.

Haemoglobin is a very efficient, but not "clever", molecule – even though it may appear so. For it not only delivers oxygen, it also removes the waste product carbon dioxide which is then carried to the lungs to be dumped outside. The more metabolically active is a cell, the larger the quantities of oxygen it consumes and of carbon dioxide it expels. When carbon dioxide binds to haemoglobin, its affinity for oxygen is reduced. Thus, the greatest delivery of oxygen automatically occurs to the busiest part of the factory. These devices are not simply beautiful in their sophistication and efficiency – they are humbling to the nth degree.

Disease

If, due to a structural imperfection, the molecules do not work properly, the organism could be in deep trouble. Sickle cell anaemia – a genetically inherited disease first recorded by the Chicago physician James Herrick in 1904 – illustrates the costly "mistakes" that can survive in genomes. The haemoglobin-bearing red blood cells have an unusual sickle shape, caused by precipitation of the haemoglobin. One consequence is that oxygen delivery to the body's cells is impaired. The sufferer may be, among other symptoms, "short of breath".

In 1949, Linus Pauling, the Nobel prize-winning chemist, and his colleagues determined the difference between standard haemoglobin and sickle cell haemoglobin. The molecular structures of the two substances are slightly different. Since haemoglobin is a protein, coded for by DNA, their finding showed that sufferers from the disease had a different sequence of DNA bases to non-sufferers. Their work was a landmark – the first to establish the molecular basis of a disease. Since that time, it has been established that sickle cell haemoglobin differs from normal haemoglobin by only two amino acids out of about 600.

Single base changes can have catastrophic effects for organisms. A single specific base change in the gene that codes for the receptor for a male sex hormone renders that receptor inactive. As a consequence, the unfortunate individual possessing this genome – which is otherwise that of a normal male – develops female rather than male sex organs, and is believed to be female at birth. The problem is only identified after puberty when the full complement of female attributes fails to develop. *One* specific base change in the entire human genome of approximately 3×10^9 bases fundamentally and tragically changes the life of an individual. As we shall see later (chapter 12), in the light of such observations, what price free will?

Disease is often based upon genetic inheritance. When this is not the case then the environment is the culprit – as widely disseminated in the case of smoking (but sadly countered by culture and advertising). But susceptibilities to environmentally-induced cancer are also influenced by genetic inheritance, for there exist genes that produce "tumour suppressing" proteins. The body has evolved strategies to reduce the probability of most cancers in the young – strategies that become less efficient with the increasing numbers of mutations that accompany advancing years.

Bacterial and viral infections are common environmental causes of disease. These agents were, prior to World War II, bigger killers of soldiers than was slaughter on the battlefield. In the Crimean War (1853-56), 16,000 British soldiers died of sickness and only 2,600 were killed in battle.[2] The evolutionary advantage to the microbe of transmission from body to body explains the symptoms that accompany the disease: "coughs and sneezes spread diseases" – whooping cough, the common cold and influenza; diarrhoea spreads cholera and the common "stomach bugs"; and genital sores promote the transmission of syphilis. The bugs find their next victim through amazing strategies.[3]

Since Linus Pauling's work – only around two generations ago – the growth of knowledge has taken us to the opposite pole: all disease has a molecular basis. But if host-specific bacteria and viruses are to survive, their lifestyle demands that they do not completely wipe out their host – the goose that lays their golden egg. The forces on the battlefield must be finely balanced, and the armies of the host must be able to fight back.

Defence

A general is just as good or just as bad as the troops under his command make him.

– General Douglas MacArthur

The human – more generally, mammalian – defence against attacking viruses, microbes, multi-cellular organisms and an enormous variety of foreign (ie non-self) molecules, is the immune system.[3] Its soldiers are more amazing than any General could reasonably demand.

Military efficiency is known to involve the loss of one's own in ways that are heart-rending to the humanist. Many British wounded were successfully evacuated from Dunkirk at the start of World War II, but a lower priority was given for the evacuation of wounded than for the fit. The sacrifice of impaired or compromised cells as part of the immune response illustrates that wars have, at different levels, been conducted in analogous ways for a long time. Social groups of cells exercised these methods long before human societies.

Cells involved in defence arise in the bone marrow – not only are bones no bigger and heavier than they need be to do their supporting work, but the hollow interiors of the larger ones are put to good use. An initially homogeneous population of cells in the marrow differentiates later into a number of cell types that react to invaders. In terms of a war analogy, civilians can be turned into troops. Upon invasion of the body by a virus, some of the invaders are eaten by one of these types of cells – macrophages. Some remnants of the meal – molecules that are characteristic of the virus surface – are subsequently displayed on the surface of the macrophage. The enemy's regimental colours are being displayed to the body's defence system.

Next, a second type of defence cell – a T cell – recognises the enemy signal displayed on the macrophage. This recognition causes the T cell population to multiply. Thus a system that recognises the attacker is amplified; or, continuing the war analogy, the government is increasing the size of the state attack/defence system. The T cells also activate the production of yet another defence – a B cell. The B cells act as factories for the mass production of molecules – antibodies – that cling to an invader, and thereby prevent it from attacking other cells. This strategy compares to the use of prisoner of war camps in human wars.

Additionally, the antibodies, while clinging to the invader, may simultaneously bind to molecules that induce chemical reactions that destroy virus-infected cells. This last strategy involves providing, to the antibody that recognises the invader, a powerful weapon to destroy it – the trained soldier has been given a gun. Since an organism is a society of cells, the destruction of virus-infected cells is analogous, in human societies, to the execution of individuals who are "infected" by enemy propaganda and can spread its destructive message.

In immunization, the surface molecules of an invader that has already been killed, or is otherwise incomplete, are presented to the body. The antibodies that recognise this surface are then formed. Because the body's defence system has a memory, it is primed and amplified in readiness for the possible future arrival of the complete, and therefore potentially dangerous, attacker. Immunization is analogous to having a large standing army, primed to know the characteristics of the enemy prior to war.

Since the body may be challenged by literally millions of different foreign invaders, each with a unique molecular shape, however does it manage to make antibodies recognise *each one*? It would be possible for each one of us to be born with say 100,000,000 genes, instructing for the construction of an equal number of antibodies. But this would be a relatively inefficient solution to the problem, requiring a vast amount of pre-programmed information. And we know it doesn't happen, for there are only around 20,000 genes for the construction of a complete human.

Evolution found an economical solution: the enormous variety of antibody molecules is constructed from a limited pool of blueprints. This apparently miraculous achievement is possible since a limited pool can provide a large number of combinations. Suppose there exist 1,000 variants of gene A, 1,000 of gene B, and 1,000 of gene C, but that we require 1,000 million antibodies. Then, through all possible combinations of the types A, B and C, 1,000 million ($1000 \times 1000 \times 1000$) different antibodies could be generated. Combination permits the generation of enormous variety. Maths, not miracles, comes to the rescue.

But Nature is so smart that it does not burden, through direct inheritance, the organism with 1,000 variants of A, B and C. Slightly different copies of genes are generated *during life*, through an unusually large mutation rate within the inherited prototypes of A, B and C. The mutation rate that is typical in the whole organism is only one change in about 10,000,000 for every cell division. The mutation rate within the genes of the immune system is one base change in about 4,000 for every cell division. This faster mutation rate generates variety more quickly, although the mechanism by which this unusually fast mutation rate is achieved is not yet known. Manfred Eigen estimated that the maximum rate of mutation permissible without loss of information by scrambling the previously evolved code is around one base per 1,000 per cell division. Thus, the fast mutation rate observed for the generation of antibody diversity is close to the maximum rate that could, in principle, do the job. The body lives within the realities imposed by scientific law.

The exquisite Darwinian nature of the immune system was made clear from the work of Caesar Milstein (Nobel Prize in Physiology and Medicine, 1984) and others. Mutations and combinatorial diversity lead to great variety. Out of all this variety, the antibody with the best fit to the invader is fittest for the task in hand and is selected for production in quantity. In finding a best fit, the body anticipated the activities of tailors. With a wide range of shapes entering the shop, it is too expensive to stock for the needs of all. Bespoke fits are provided by combining tailor-made bits and pieces. But the tailor of the immune system is not honed to satisfy the shape that has entered the shop; rather it is the Sweeney Todd of the tailoring world with the aim of executing the customer.

But the bugs and viruses often meet the challenge, as they must do if they are to survive. Among the tricks they employ is that of changing the molecules of their surfaces – their suits – again through a relatively high mutation rate. The immune system may then be too slow in recognising a novel invader, with fatal consequences – which is why researchers have to guess what might be the best vaccine for the next flu epidemic. The HIV virus gets up to the same kind of trickery, so drugs that are successful in early treatments may fail a patient tomorrow. The pharmaceutical industry will never be able to relax.

Through the intertwining effects of genes and environment, pathogens and reluctant hosts are mutually modified. And "mind" itself is the product of a molecular structure arising from the convolution of genes and environment.

Mind

A patient, when presented with a rose and asked to describe it, commented, 'About six inches in length – a convoluted red form with a linear green attachment.'

– Comment by Oliver Sacks on one of his patients, in his book
'The Man who Mistook his Wife for a Hat' (1985)[4]

During evolution, in complex animals there was selection of a vast assembly of neurons – a brain. The brain allowed information not only to be rapidly sensed, stored and processed but also allowed rapid responses to the incoming information and its processed product. It evolved at the front end of the animal because it is a good thing to sense, and react rapidly to, the environment that is being approached.

Mind is usefully defined as the outcome of the functions of the brain. It is a measure of the complexity of mind that, in Oliver Sacks's account, the description "rose" was erased from a brain in which much analytical precision remained.[4] It was the same patient whose behaviour gave rise to the title of the book, for on one occasion while searching for his hat, he "took hold of his wife's head, tried to lift it off, to put it on". Yet in many other respects, this patient's abilities were normal, and his musical appreciation and performance were excellent. The problem of this patient is readily accepted as "a disease state". But, owing to our competitive natures and the demands of society, we are less likely to sympathise with much more subtle variations in brain function – those that make a person "arrogant", "stupid", "selfish", "boring", etc. And usually praise those that lack these conditions.

Knowledge of the brain is very limited. All the way from fish to mammals, three sub-divisions (hind-, mid- and fore-brain) can be discerned. The hind-brain lies nearest to the top of the spine and is continuous with the spinal cord. The hind-brain controls the automated coordination of body movements – as in walking, the beating of the heart, and breathing. It has ancient origins, and is heavily implicated in ancient functions. Hence the headless chicken retains some primitive autonomic function: it may still run. Moving further away from the top of the spine, the mid-brain follows the hind-brain. The fore-brain – notably large in humans – extends from the mid-brain.

The mid-brain regulates sleep and responses to sound, suggesting that it has also evolved from an ancient structure – but not as primitive as the hind-brain. The fore-brain shows all the signs of incorporating "newer parts" – including consciousness, and the emotions – instincts of which we are aware when calm, but exercised with reduced control when aroused. The successive structures appear to have been extended from the primitive brain stem. Inevitably, there is some hazard in ascribing functions solely to one of the three main sub-divisions, for there are interactions between them; as when an individual, at one instant sedate, later becomes part of a primitive mob. Hardly surprisingly, *Homo sapiens* show some apparently contradictory traits.

The human brain weighs three to four pounds (1.3 to 1.8 kg), and contains about a hundred billion nerve cells. The average weight of one of these neurons is about one hundred millionth of a gram – far less than the weight of a piece of gossamer that can be seen with the naked eye. The neurons are thin and branched cells, with a thicker bulb evident within their structure. Information is passed along their length in the form of short electrical impulses, lasting about one thousandth of a second. Since any one neuron makes many connections to others, the total number of connections in the brain is greater than 10^{13}.

Signals from one nerve cell to another must cross inter-nerve gaps – synapses – around a millionth of a centimetre wide. Small molecules – neurotransmitters – diffuse across this gap to transmit the signals. When they arrive at the second neuron from the first, they bind to a receptor. When drugs that can competitively bind to this receptor enter the brain, *via* the bloodstream, behaviour is modified. Brain function is the outcome of chemistry.

Eric Kandel was awarded a Nobel Prize in 2000 for his discovery that synapses are modified in learning and memory. He used the nervous system of a sea slug (*Aplysia*) that has relatively few nerve cells – about 20,000. He showed that the chemical modification of proteins – adding phosphates to them – in the borders of synapses is involved in the generation of short-term memory, which lasts from minutes to hours.

More powerful and long-lasting stimuli result in long-term memory – remaining with the slug for weeks. In contrast to short-term memory, which requires the modification of *existing* proteins, long-term memory requires the synthesis of *new* proteins. This new protein synthesis leads to alterations in shape and function of the synapse. Kandel also carried out studies in mice, and showed that the same type of long-term changes of synaptic function also applies to mammals. Memory and mind work by changing the structures of molecules,[5] through operation of the principles of physics, chemistry and biochemistry.

The sharp division of memory into categories of "short-term" and "long-term" may seem rather stark, but is supported by common experience of our ageing loved ones. And is sometimes manifested to the extreme, as in a case cited by Oliver Sacks:[4] a patient showed a complete inability to remember any current experience beyond about a minute, but could recall experiences that had occurred more than thirty years previously. The lack of short-term memory was so severe that the patient could meet someone who would then leave the room, only to return a few moments later and be treated like a complete stranger. In 1975, *nothing* could be remembered since 1945, even to the extent that the patient imagined himself to be 19 at age about 45 – and would temporarily turn ashen when confronted with his own greying features in a mirror.[4] One molecular mechanism has failed; another continues to function.

Environmental influences constantly modify the brain because these stimuli affect the expression of its hardware. In an experiment carried out by Jenkins, Merzenich and colleagues,[6] a monkey was induced to carry out a task using only the centre three fingers of its hand. After the task was repeated several thousand times, the area of the brain receiving signals from these three fingers was expanded at the expense of that which would otherwise have been devoted to the other fingers. The same effect operates in violinists.

The most perplexing aspect of brain function is consciousness; for example, our ability to conjure up the face of an absent friend in our "mind's eye", or to think about a problem. Francis Crick related a discussion in which he posed the question: "What is consciousness?" The response was: "It's like having a television set inside your head." To which he responded, "But who is watching it?" Advances in the neurosciences provide a framework.[7] It is proposed that, when we conjure up the face of X in our mind, a set of interconnected neurons which encode that image are induced to "fire" – to transmit electrical impulses within neurons, and chemical messengers between them. Thus, even when the cells associated with our five senses are not being stimulated, we are able to recall faces and think about problems and experiences – internal processing going on in the absence of an input from the external environment.

When information transmitted from our immediate environment is processed, the brain responds in a manner that is more detailed than can be derived from this input. For example, the image of a house is constructed *via* photons passing from the building, through a focusing eye lens, to an essentially two-dimensional retina at the back of the eye. Yet from this essentially two-dimensional information, the brain infers a great deal about the distance, height, depth and width of the house. To do this, it relies on previous experience – houses are normally large, so that a small image implies a large distance – a programming successfully exploited in movie studios.

It was commonly believed in the past – and it is still assumed by many in the present – that mind might be separate from body, and that this separation provided a basis for an "afterlife". The conscious reflections of the human brain upon dying and death are horrifying. The "afterlife" proposal was therefore inevitable. But, through experiment, we know now that these presumptions are false – memory is coded in molecules. Mind is a part of body; thoughts and consciousness derive from the working of cells. The idea of an afterlife is wishful thinking for life functions through earthly cells; cells that can now be manipulated.

Mutations more commonly compromise survival rather than promote it. Scientific discoveries now allow some of adverse mutations to be erased in a humane manner. Although an anathema to many, there are sometimes good reasons "to interfere with Nature".

But even when there are opportunities to improve the quality of life, it is common parlance to speak of "human perturbation of the order of the natural world". But humanity is part of the natural world; vividly illustrated by the sharing of around 1,000 genes by all organisms; and by the sharing by fruit flies (14,000 genes) and humans (around 20,000 genes) of no less than 70% of equivalent genes. What is described as "interfering with Nature" is no such thing: the activities of humans evolve by natural means and are a part of Nature. So, when human activities change the globe, this change occurs by natural means. A large brain and the harnessing of technologies have made the human impact on the environment relatively large, but that impact remains "natural". Cities with superhighways evolved as naturally as did the nests of termites.

When two species interact strongly over many millions of years, their forms are reciprocally manipulated through the interactions. Flowers and pollinating insects co-evolved, each becoming increasingly reliant upon the other;[8] bees, and the flowers they visit, mutually adapted to a degree that astonishes the careful observer – so sophisticated that orchids of the genus *Ophrys fusca* seduce a male bee. The insect attempts copulation with the flower in which a central portion bears a striking resemblance to the head and front legs of an insect. Even where the resemblance to legs is absent, an amazing resemblance to an insect head and eyes is seen in the Thai orchid.

In the breeding of plants for food crops, and of dogs and pigeons, humans applied severe pressures; large changes were induced within centuries or millennia. Organisms are moulded and manipulated through myriad environmental pressures. But, if it is known what the plans are that make the organism, how the plans are used to make its devices, and what the functions are of these devices, it becomes possible to make and manipulate them in new ways. And humans are certainly busy manipulating – new and intense environmental pressures.

Mechanical Tinkering

In a car, malfunction may occur because a device is unable to get a product to another device – say, get petrol to the spark plug should a fuel line be blocked. In a similar way, the human systems for reproduction often have shortcomings. So, tinker to help them. Multiple ovulations are induced, several eggs then fertilised outside the body (IVF), and the fertilised eggs then implanted in the womb. Or sperm, which are prevented from penetrating the membrane surrounding an egg due to deficient molecular structures, may be directly injected into eggs (ICSI). Couples, formerly unable to conceive, have children.

Sometimes the fertilised eggs are implanted in a surrogate mother. Using the devices of other humans causes ethical problems. But since ethics are applied less in animal welfare, this procedure is common in cattle; genes of good milk or beef producers selected for transmission at the expense of the genes of surrogate mothers. The evolution of cattle put on to a new track.

Genetic Screening

Whether an animal develops through traditional, or mechanically aided, fertilisation, there is the possibility of genetic-based disease. To use genetics to cure these diseases requires knowledge of the DNA that is causing the problem – the genes must be screened. Genetic screening is the analysis of the DNA of an embryo, foetus or adult. It informs about the probability, or even certainty, of certain future crippling or fatal diseases.

A guiding principle regarding the use of genetic screening, expressed in *Time* magazine in 1989, was: "Information about people's genetic constitution should be used only to inform and never to harm." The sentiment is admirable; the difficulties clear. DNA analysis can identify those who later in life will suffer from certain genetic diseases. If a genetic analysis establishes the presence of a specific aberrant gene then this information can be beneficial if useful action can then be taken (eg breast cancer). But if an invariably fatal gene is spotted, the information is likely to ruin what otherwise might have been a period of contentment: to "inform" has been to "harm". The screening genie is out of the bottle and, if asked to perform, may provide contentment or anxiety.

Genetic Engineering

If a faulty gene has been identified through genetic screening, genetic engineers have a chance to exercise their skills. In some cases, the desired genes are synthesised in the laboratory; these bits of DNA string can then be tied into the long DNA strings of organisms. In other cases, DNA cut from one organism is inserted into another species. The cutting and tying processes are both carried out by enzymes.

Startling effects come from genetic engineering. The DNA of a mouse may be manipulated such that a gene that codes for growth hormone is over-expressed; the result: very large mice. It stuns that a single molecule causes the production of countless billions of cells – whether of rodents or humans – to be increased in exactly the required proportions.

Beneficial effects come from genetic engineering. Mammalian genes – say a pig gene coding for insulin – can be transferred to, and expressed in, bacteria. Using standard large-scale technology for the mass-production of bacteria, pig insulin is then mass-produced. Pig insulin differs in only one amino acid from human insulin, and so the former is converted into the latter before human use. Despite this conversion, a Rabbi may be consulted with regard to the ethics of providing a Jewish diabetic with human insulin derived from the insulin gene of a pig. Irrespective of the response of the Rabbi, positive or negative, the conclusions of science, and the requirements of culture, are different: science concludes the two products are the same; culture leads adherents to a particular faith to consider whether the package from which the gene came has relevance.

The above solution to a medical problem is still a rather messy business – the gene expression is used to manufacture a specific protein or peptide which must then be separated from other stuff, transported and finally delivered to the body by injection. Exactly as, when a faulty car has been delivered, the solution is to manufacture properly the faulty component, separate it from all the other stuff in the factory and send it to a garage where a local expert puts it in the correct place. Is there a neater solution?

Gene Therapy

The above rather messy business could be obviated if the car makers could provide a plan that, once placed within the car, would automatically manufacture, and put in the appropriate place, the perfect component to replace the defective one. Car manufacturers cannot provide such automated internal repair.

But parents may be able to provide, to a suffering child, the means for automated internal repair. If their child suffers from an almost certainly fatal genetic disease, they may choose to have a further child. When the fertilised egg has divided to a collection of a few cells, pre-implantation genetic screening is performed. Embryos possessing the gene that can save the life of the existing child, and tested to ensure an immunological tissue match (coded in HLA genes on chromosome 6) with that child, are implanted. Cell transplantation from the younger to the older sibling can then remove the threat of an otherwise often fatal illness. This procedure allowed Molly Nash, at the age of six, to successfully receive an umbilical blood transplant from her baby brother Adam in 2000.[9,10] Critics speak of "saviour siblings", but Mr and Mrs Adams spoke of loving both their children, and of the bond between them.

Gene therapy can also be achieved by specific gene implantation in the laboratory. Immune deficiencies, and many other genetically based diseases, are so serious that children suffering from them are condemned to an early death. Cells can be taken from the bone marrow of the child, the appropriate gene inserted, and the "repaired" cells then infused back into the child. By 2005, this technique was saving lives without signs of adverse effects.

There is a third way in principle to carry out gene therapy. If a disease-causing gene could be replaced by a good copy in a fertilized egg then not only the developed child/adult would benefit but the gene would also be passed on to subsequent generations – germ-line therapy. Descendants would not suffer from the disease caused by the tainted device. But the long-term effects of providing genes in this way to humans are unknown; the hazards particularly great since germ-line therapy has the big disadvantage that genes are usually integrated randomly and may well damage the genome at the point of integration. It is also an incendiary suggestion, having a whiff of eugenics and fascist dictators.

In any of the methods for gene therapy, there is another limitation of its power as a cure-all. By 2009, there were indications that many genetically based diseases involve large numbers of genes. Transferring one or even a few genes was looking somewhat less promising as a cure-all.

Cloning Animals

Germ-line therapy of humans is regarded as unethical and dangerous for the reasons given above. But ethical considerations are diluted in the manipulation of cattle. Commercially more useful bulls are produced from only one production line. They are subject to the same kind of evolutionary pressures that led in the 1990s to the demise of the Trabant, but the survival of the lines – with continuing improvements – for Mercedes. The organism equivalent of this outcome is to clone, and to keep selecting, what works best.

Cloning refers to the production of a genetic replica of a cell, tissue or organism. It is most famous at the level of organisms. The DNA of cells is contained in a nucleus. If the DNA is removed from the egg cell of a frog and then replaced by the DNA from the gut cell of another (adult) frog then the modified egg cell produces a new frog. This frog is a clone of the frog that provided the gut cell. John Gurdon of Cambridge University reported this result in the early 1960s.

However, it was the cloning of the sheep Dolly in 1997 that caught the attention of the media – and hence of the world. Using the same method as for the cloning of a frog, the DNA from a mammary gland cell – hence Dolly – of an adult sheep was directed to produce an embryo, and this embryo was implanted into a surrogate mother. Dolly was a clone of the adult from which the mammary gland cell was taken. Cloning is used to produce "high yielding" dairy cattle in the United States.

Cloning Cells from Embryos

The cloning of complete humans is beyond the pale. But the cloning of selected human cells is being considered. In this respect, the cells of very early embryos are of huge interest for they include stem cells – cells sufficiently plastic to produce any of the approximately 220 types of cell found in the adult.

Perhaps these stem cells are the fount with the greatest potential to cure human disease? For, if such a cell could be taken from an early embryo and maintained in culture, it would be available to treat the adult that developed from the embryo should she/he suffer from a genetic disease that had been induced by the environment. This treatment would be possible without rejection by the immune system. Diabetes and Parkinson's disease appear to have two components: perhaps genetic predisposition, but added risk from environmental effects on genes during life. So, both diseases – in which specific cell types function abnormally – are candidates for treatment in this way if the genetic damage is environmental.

Very early human embryos consist of a ball of about 100 cells, reached about 4 days after fertilisation. They are available in limited supply from fertility clinics. On the one hand, they are often discarded, and may provide human embryonic stem cells without destruction of the embryo.[11] On the other hand, since embryos have the potential to develop into humans, the cloning of these cells to obtain their plastic products is also a contentious area of research. Therefore, it is better perhaps to seek to provide them another source. Could cells of the adult be reverted to their womb-like state? The cloning of organisms provided a clue.

Cloning Cells from Adults

It is the proteins that surround the DNA in the egg that cause it to start the building of an organism from scratch. In animal cloning, these egg proteins reset the genetic clock, for it is the environment of the genome that determines which genes will be "switched on" and which will be "switched off". The successful cloning of frogs and sheep attests to this remarkable way in which the genetic plan can be made to start again: from "time zero". Thus, what may have seemed like a fantasy goal – like asking the car manufacturer to miraculously convert the ancient banger back to a pristine new vehicle – was apparently achieved though cloning.

Therefore, embryonic stem cells can in principle be generated by taking the nuclear DNA from a cell of an adult patient. This "patient DNA" can then be implanted into a human egg from which the original DNA has been removed. The egg proteins function to reset the genetic programme of the patient DNA to "time zero". Egg cells modified in this way can then undergo a few divisions to provide embryonic stem cells.

But human eggs are in limited supply. Therefore, efforts have been made to modify, without the use of human eggs, cells from human adults so that they would once more behave like stem cells – to reverse their paths back towards t = 0. Researchers searched for a change in gene expression that would achieve this goal. And by introducing only four genes, human skin cells were modified so that they had the plasticity to produce a number of cell types.[12] The possible use of such cell types in any required future treatment has an attraction: since "the plans" of the patient would be used, there are no problems with rejection by the immune system. But, as often, there are counter points, for the adult skin cells will already contain adverse mutations due to the environmental insults experienced during the lifetime of the adult. These stem cells have memory of being old at heart.

Not surprisingly, methods of manipulating the human genome are topics of controversy. Collectively, the techniques are criticised by some as "playing God". When James Watson was asked to comment on this criticism, his response was "If we don't play God, who will?" A decision not to develop a ball of cells proven to carry genes for tragically severe disease is a sure way to reduce their transmission; to act against the prevalence of the disease. The application of new knowledge, with due consideration of ethics, can be better than the suffering that sometimes comes with random variation.

Tests of truth contribute to ethics. But ethics lie in a field where the environment of a lifetime profoundly affects what we believe – often too complex to be examined in this light of "truth". But how can we be guided towards truths? What predictions are to be believed, which ignored? When can we reliably know "what will happen next"?

10

Predictions and Feedback

Boring Predictions

The frequencies of various kinds of electromagnetic waves are known and correlate with their properties. X-rays vibrate around 10^{18} times per second; pass through flesh, but scarcely through bone. Light waves vibrate about 10^{15} times per second; pass quite efficiently through air, but not through flesh. Radio waves vibrate about 10^6 times per second; pass through thin house walls, but not through the rock walls of a deep tunnel.

The frequencies are correlated with other properties of the radiations, making reliable predictions possible. But, in a sense, they barely qualify as "predictions". Better regarded as facts that may bore; barely more impressive than the prediction "the Sun will rise tomorrow"; important only as knowledge upon which to build. Making other predictions poses more of a challenge.

What Happens Next?

That Herr Hitler...should be given the chance of showing that he is something more than an orator and an agitator was always desirable.

– 'The Times', London, 31 January 1933

Making successful predictions is very difficult – especially when they are concerned with what will happen in the future.

– Niels Bohr

Richard Feynman was among the most influential physicists of the second half of the twentieth century. From his base at Cal Tech in Pasadena, he delivered the dictum: "Given a known set of starting conditions, saying what will happen next is what physics is about." To this man, intellectual puzzles were a challenge that could not be ducked.[2] In his days at a centre for US Government sponsored physics at Los Alamos in New Mexico, he was known for his ability to pick locks and crack security codes; not only for the challenge but also to cause consternation to bureaucrats.

He pointed out amazing aspects of waves. We sit in a room pervaded by radio waves from Moscow, evidenced by hearing a baritone's voice from that city; simultaneously receive a television signal from Los Angeles; hold a conversation with a second party, yet eavesdrop on the conversation of a third. All these waves in one room, but all deciphered without being hopelessly mixed up! Awesomely, there is a deconvolution of numerous waves by the brain – astonishingly coded for in the DNA of miniscule sperm and egg. And all emanated from a cauldron of indescribably hot particles given few rules, the essence of which have been described in around 200 pages.

Feynman stretched the amazement by using the analogy of a swimming pool with lots of different waves on its surface caused by lots of people jumping in. To paraphrase him, "It is as though the vibrations of a fly, floating on the surface of the pool and thereby experiencing the waves, would allow us immediately to know exactly when and where everybody jumped in!"

Applying Feynman's dictum "What happens next?" to the fly requires a complete analysis of all its motions as a function of time. Such calculations are formidable because they involve a large number of variables. The remarkably successful predictions that have come from many areas of physics are possible because the number of variables is small; chemistry less successful because it often deals with more complex systems; biology successful to an even smaller degree because its systems commonly involve even more variables; sociology only partially a science.

So, one had better be careful in making predictions about the future behaviour of complex biological systems. By 1968, in countries with advanced technology, infectious diseases had declined dramatically – thanks to antibiotics, vaccines and improved hygiene. So, a bold Surgeon General of the United States pronounced: "*The time has come to close the book on infectious diseases.*" He was unaware of the difficulty of reliable prediction for very complex systems.

Despite Bohr's maxim, individuals often make bold predictions regarding the future of very complex systems. Be they military leaders, politicians, chief executives, investment bankers, social commentators or media-with-a-mission types, they may gain high status by doing so – one reason why it is done. More frequently than not, it transpires in the long run that they have feet of clay. For their world is far too complex. It is awash with data, with no immediate test to ascertain whether truth or false; even where true, only a small fraction of it affording useful information. We are rarely equipped to solve complex problems.

But trying new combinations, testing for fitness of understanding and function and selecting accordingly is the paradigm that has worked well. Individuals who dramatically advanced our world observed carefully; tried combinations that were new, tested for fitness through calculation or experiment; and so discovered principles, or devices, that were selected for transmission.

Changing the World

Nothing is too wonderful to be true if it be consistent with the laws of nature.

I am busy just now again on electro-magnetism…

– Michael Faraday

Newton, Darwin, Maxwell and Einstein changed the world. Others built upon steps taken by less famous figures.

The fast transmission of nerve signals by an electric current was an understanding only slowly accessed: initiated by an observation in 1771 by the Italian anatomist Luigi Galvani. He observed that the muscles of dead frogs twitch when struck by a spark of electricity – hence "galvanised into action". The Italian physicist Alessandro Volta (1745-1827), after whom the volt – the "driving force" for the passage of an electric current through a conducting wire – is named, developed the principle of the battery. Humans became able to store electricity – an asset that could be rapidly transmitted.

Electric discoveries were a speciality of the British scientist Michael Faraday (1791-1867).[1] He gave one end of a current-carrying wire, located in a magnetic field, freedom to move by immersing it in a bowl of mercury. The wire rotated – mechanical work could be obtained from electricity. *Ergo*, electric motors – in cars, hair dryers, refrigerators, power mixers and drills. Conversely, he also discovered that if mechanical work is done to move a conducting wire through a magnetic field, an electric current is generated in the wire. *Ergo*, the generators of cars, and electricity power stations. Passage of an electric current through a wire of appreciable resistance raises its temperature. Capitalizing on this effect, the American genius of technology Thomas Alva Edison (1847-1931) invented the electric light bulb. Booms followed these discoveries, and the world was changed.

Further developments in the understanding of electricity and magnetism were to revolutionise society in yet another way in the twentieth century. Following the description of light as a combination of electric and magnetic fields by James Clerk Maxwell (chapter 1), it was demonstrated that these fields fluctuated at very high frequencies – around 500 thousand billion times per second. Moreover, it became apparent that light constituted only one small part of a wide spectrum of "radiations" – extending from cosmic rays and X-rays at one extreme to radio waves at the other. All these radiations travel at the speed to light. Hence, if information could be encoded into them, it could also be transmitted at the speed of light. Radio waves were successfully "modulated" to carry the additional information. The birth of radio and television became possible.

Another chance to change the world derived from the work of the already familiar Leibniz. As the conventional number 1679 is read from right to left, it tells there are 9 ones, 7 tens, 6 hundreds and 1 thousand. A system of counting based upon the ten numbers 0-9; evolved from the ten digits of the hands. But Leibniz showed around 1679, and published in 1701, that it is possible to encode any number based on only two numbers: the binary system, using only the digits 0 and 1. Since the binary system has no number "2", a move to the left is required to count even two objects – represented as 10. Further moves to the left encode the numbers of fours (100) and eights (1000). An expanded list of conversions is:

Base ten:	256	128	64	32	6	8	4	2	1
Base two:	10000000	10000000	1000000	100000	10000	1000	100	10	1

Thus, 15 in the binary system is 1111; 57 is 111001; 1679 is 11010001111.

The binary way of counting has an advantage: through having large numbers of electrical switches that are either simply in the "on" (1) or "off" (0) positions, any number, or indeed any kind of information, can be encoded – the basis of the modern computer.

Thus, information came to be transmitted not only at the speed of light but also then processed in, and displayed on, binary-based computers. With the advent of solid-state chips that encode large amounts of information in a tiny space, cellphones that transmit and receive tunes and photographic images became available. Satellites were put into orbit through fast chemistry – rocket engines – and an understanding of the laws of motion. In consequence, sports events are almost instantaneously transmitted with essentially perfect fidelity across the globe. Widely enjoyed; rarely with wonder.

It is told that Faraday, when explaining one of his discoveries to William

Gladstone, the then Chancellor of the Exchequer, was asked, "But, after all, of what use is it?" Faraday allegedly answered: "Why, Prime Minister, someday you can tax it." This anecdote has a moral: key discoveries are not normally made to bureaucratic instruction. Rather, they are made by unfettered and imaginative enquiry by deep thinkers; thinkers not encumbered by conventional wisdom, rigid system of reports and timelines.

Nerves and electric eels, perhaps inevitably, evolved useful devices using electricity long before humans. After humans had stumbled upon electricity, no one sitting by candlelight in 1800 could have predicted the subsequent revolutions. But the message is clear: if a nation is to lead in science and technology, it must avoid rigidity. Educational systems promoting wonder, enquiry and rationality are the way forward. And, despite the limited adoption of these criteria by some nations, it was only with the coming of the second half of the twentieth century that the question "Why boom and bust?" could, in its simplest forms, be quantitatively addressed.

Fish and Finance

Elegance is not optional.

– Richard A O'Keefe in 'The Craft of Prolog'

Fire a shell from a gun at a known velocity and at a known angle. If there is no variable wind, a physicist/engineer predicts with impressive accuracy where it will land. This system is "linear". There are known relationships between the variables – distance, velocity, acceleration and time – and these do not change as the experiment proceeds. But some systems involve "feedback" – where changes occurring at an early stage of the game influence the rate of changes that are yet to come.

Positive feedback clearly played roles in the rapid utilisation of electrical phenomena in the last 200 years. One discovery followed close on the heels of another, and there was an increase in the rate of change. Until the second half of the twentieth century, physics and mathematics did not provide an understanding of systems involving "feedback" – also described as "non-linear".

Even apparently simple systems that involve feedback show surprisingly complex behaviour. Smoke rises from a stationary cigarette in a room where the air appears perfectly still. It may initially rise in a relatively straight and unperturbed way, but then suddenly pass into a violent swirling pattern. Water at one instant passes smoothly down a stream, but then the flow suddenly becomes turbulent. Why? Our understanding of such phenomena has come from a few "scribblers": thoughtful types willing to put down an equation on the back of an envelope to try to figure out what is going on in a field now known as the study of chaos.

Here, the approach taken by Robert May in the early 1970s is followed.[3] Robert May is Australian by birth but pursued his later career in the UK, eventually to become Lord May and President of the Royal Society. His approach to lectures, and frankness, leads to delightful outcomes in a country that has many formal traditions. At the beginning of the twenty-first century, "elitism" is a dirty word in the thoughts of many. When giving evidence to a Parliamentary Select Committee on the activities of the Royal Society – the UK's foremost scientific society – he was faced with a political trap: asked if the Society was elitist. But he was up to the challenge: "Yes, in the same way as the England football team." If you want the best, you select the best. The problem for society lies in the development of strategies that optimise the numbers appropriately described as "the best".

May tells the story that, when teaching physics at the University of Sydney, he would stand on a chair at one side of the lecture theatre while holding a heavy metal ball close to his groin. The ball was suspended from a wire attached to the centre of the ceiling of the lecture theatre. So, when the ball was released, it swung in a pendulum motion towards the other side of the lecture theatre, from where it then returned towards him with gathering speed. Undergraduates feared for the future of his manhood. But May was of course demonstrating, with confidence, that the laws of physics – in this system, without feedback – made a reliable prediction. The motion of a pendulum is damped and so, given only the considerations of Newtonian physics, the ball must stop a few centimetres from its apparent target.

Robert May made important contributions to the understanding of what is going on in "chaotic" systems – specifically systems with negative feedback. He used an equation that is so simple that anybody with one year of algebra and a calculator can check it out. The equation was applied to calculate the population of some organism – say fish – in a particular niche – say a pond – with the passage of time. Do not be put off by "fishes" and "equation", for an understanding of why stock markets crash emerges from a study of negative feedback; future wealth determined by taking a little time.

A simple starting point assumes that the population is determined by only two variables. First, the breeding rate, for the population will tend to increase more rapidly the higher the breeding rate of the fish. The breeding rate is defined as 2 if the population increases by a factor of 2 each year in the absence of any other influence. So if the initial population were 10 fish, the next year the population would be 20, the year after

40, the year after 80, and so on. More generally, if the initial population were x fish, and the reproduction factor r, the population in the following year (x_{next}) would be x times r (written x.r). Moreover, we can keep the year-on-year numbers small and easy to compare by taking the highest conceivable population as 1 (and extinction of the population obviously as x = 0). Thus, x will always lie within the boundaries of 0 and 1.

The second variable is the limitation of the food supply: for if the food supply is limited – as with a burgeoning population, it eventually must be – the number of fish must at some point decline. The approach must also recognise that the higher the population, the larger will be the negative effect of a limited food supply. How can this effect be quantified? Within the possible population range 0 to 1, consider ones that have increased to 0.9, or in a second case 0.8. In the case of x = 0.9, this larger population is going to suffer more from a limited food supply than when x = 0.8. A larger reducing effect – larger negative feedback – is required when x = 0.9. This requirement is easily met: simply multiply x by a "reducing factor" of (1-x), which gives for the next year population (x_{next}):

$$x_{next} = x.r(1-x)$$

When x = 0.9, the new population will be $0.9 \times 2.0 \times (1.0 - 0.9)$, ie 0.18. When x = 0.8, the new population will be $0.8 \times 2.0 \times (1.0 - 0.8)$, ie 0.32. The model nicely encapsulates the fact that the higher the current year population, the more drastic will be the effect of food limitation on the following year population.

Start with a fairly small initial population (x = 0.1), and calculate the year-on-year populations – Table 10-1.

Table 10-1

Year	x	x_{next}
1	0.1	0.18
2	0.18	0.295
3	0.295	0.416
4	0.416	0.486
5	0.486	0.499
6	0.499	0.500
7	0.500	0.500
8	0.500	0.500

The population increases gradually at first and, within a few years, reaches an equilibrium value (0.5). The equilibrium value is halfway between extinction and the maximum conceivable population. The balance between opposing effects – breeding and food acquisition – determines the population of the fishes. And in the general case of only two well-defined variables behaving in this way, of all other organisms, *Homo sapiens* take note!

Now make the fish more successful in their sexual reproduction by increasing r from 2.0 to 2.8. The results are plotted as a graph (Figure 10-1).

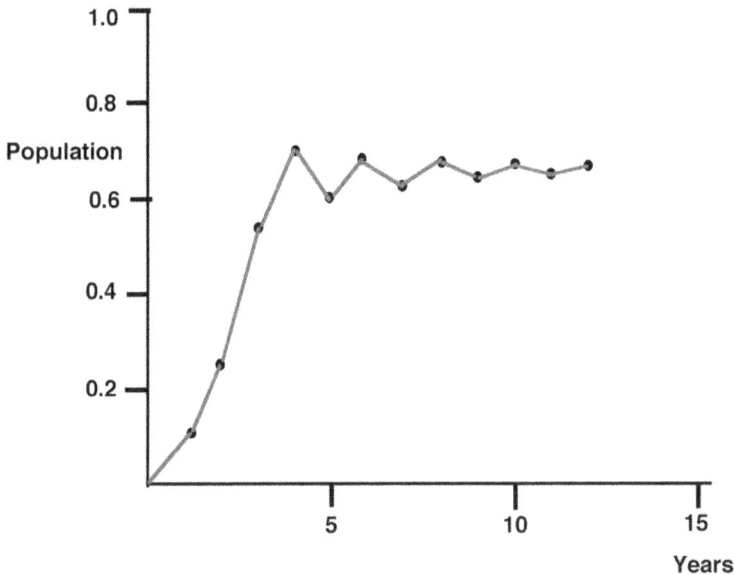

Figure 10-1. Oscillation of a fish population.

Again, the population gradually stabilises. But, in even-numbered years, it overshoots slightly; and, in odd-numbered years, undershoots slightly, before equilibrating after about 30 years at 0.643. This is because when the population is significantly below the equilibrium level, the high breeding factor helps to push the population above the equilibrium level in the following year, only for the price to be paid – in the form of the limited food supply – in the following year. There is no free lunch. But the breeding factor is not so high as to preclude the setting up of equilibrium after a few years.

Next, let the organism breed "like rabbits" – increase the successful birth rate by putting the breeding factor up to 4.0 (r = 4.0). The results – not shown, but readily calculable by the reader – are dramatic. The year-on-year populations now undergo very large changes, and without a repeat over any year. We have entered the state known as "chaos" – an unfortunate name since the outcome, far from being chaotic, is determined by the laws of science.

In the "chaos" region, even a minute change in the initial assumptions can drastically affect the outcome. For example, we may not know the value of an input parameter within 0.1%, and yet smaller variations in this parameter may drastically affect the outcome. There is a *profound* message in this example – reliable prediction is impossible. The point that the input determines the output must not be lost. The problem is that minute changes in the input can prevent us from predicting that output.

In sum, feedback control can, on the one hand, give rise to wonderful stability – Table 10-1 and Figure 10-1. On the other hand, if strongly opposing variables are thrown against each other, very large differences in the outcome can occur. To illustrate these points, the effect of widely differing breeding factors (r) upon the fish population can be calculated (Figure 10-2). When r is in the range 2 to slightly less than 3, a *single* stable population results. In the range 3 to 3.4, the population *alternates* between 2 levels (in Figure 10-2, these two levels are shown, but the alternations between the two levels are not).

At r = 3.5, it alternates between 4 levels; at r = 3.6 between 8 levels. Turn up r slightly more and soon the calculated population visits an enormously large number of values, but all lying in a defined range. The message: if two opposing effects work strongly against each other – one towards "boom" and the other towards "bust" – the outcome may be incredibly sensitive to the input. *Big* changes come with a *high frequency*. Are the stock market analysts, many of whom ignore the relevance of mathematics to their field, taking an interest?

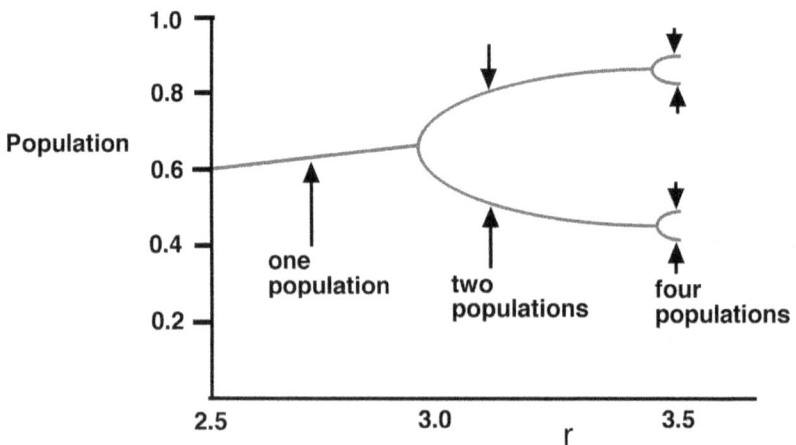

Figure 10-2. Patterns of population development.

The above considerations are far from esoteric; repeated patterns and complexity emerge spontaneously from simple rules – ones that are relevant from guppies to goats, and from fish to finance. Feedback is crucial in economics, where demand feeds back on the production level for a given product. The dangerous side of the game is that the more quickly some quantity spurts, the more rapid and severe can be the ensuing constraint – negative feedback. The consequence is large oscillations in values. Worse still, chaos. Sounds relevant to the circumstances of hundreds of millions of workers, and of banks and stock markets.

The concept of human wealth is that an asset is a marketable commodity, the value of which depends on current expectations. If these expectations collapse, the feedback effect is severe – the 1929 Wall Street Crash and the crash of 2008. When recessions and/or depressions ensue, the monetary authorities may print more money – a valid strategy if replacing only in part the notional wealth that disappeared through prior exuberance. If the money supply is too drastically increased, there follows chronic inflation: Zimbabwe at the beginning of the twenty-first century; the Weimar Republic in the twentieth.

In human societies, feedback is ubiquitous, but typically not anticipated by those exercising power. When the polarization of wealth becomes grotesque, the feedback may be revolution: France at the end of the eighteenth century; Russia at the beginning of the twentieth. Wisdom may allow some to spot when expectations move far from reasonable equilibrium values. But almost nobody dares to be a prophet of doom amid the profits of boom. Economists may speak of "rational markets"; the markets are often anything but. Start the clock in a modestly rising market. People perceive this and buy shares. Confidence is rising, people and institutions happily borrow to invest and consume, and a boom begins – positive feedback.

But in the long run we have "a zero sum game". Many greedy individuals, having seen their assets appreciate in the early part of the cycle, become acquisitive beyond the bounds of reason. They borrow on the basis of their present assets, on the assumption that prices will continue to rise. But there comes a time at which the market has priced itself above reasonable expectations. Negative feedback sets in, prices fall, debts ensue, and recession or depression follows.

Some investment professionals speak airily of " market globalisation" as though it were a phenomenon of sophisticated analysis. Parts of it are anything but sophisticated; rather "me-too" fears and greed. All-we-like-sheep behaviour is illustrated by the behaviour of the Nasdaq and Dow Jones indices on most trading days. See for example

http://www.bbc.co.uk/news/business/market_data/overview/default.
stm. The almost identical changes in the values of the indices as the day
progresses are striking. Close correlation of the two indices is common.
Perhaps initially surprisingly so, since the Nasdaq index is heavily
weighted with technology stocks, whereas the Dow Jones is constituted
from a number of "heavyweight" industrial companies. Is it not curious
that these indices of very different stocks behave, even with a resolution of
a few minutes, almost identically?

Well no, not if the hour-by-hour determinant of prices is primarily
gossip across the floor of the exchange and across the global ether. On
the day, human hopes, greed and fears dominate a long-term overview
of economic fundamentals according to sector. As noted by one of my
colleagues, the distinguished mathematical economist Frank Hahn,
"Shares do what people think they will do." There is often little that
is rational in the herd's conventional wisdom, and human opinion is
notoriously sensitive to feedback, either positive or negative. So much so
that the UK FTSE 100, given a time adjustment for longitude difference
and a small "communication time lag", often closely tracks both Dow and
Nasdaq. Independence of mind is rare. For social systems encourage, and
reward, conformity in the short-term, but often punish it in the long run.

So, in addressing the question "What will happen next?", neither
scientists nor economists can tread confidently in the area of economics.
There are too many input parameters – particularly human behaviour –
whose values cannot be well defined; which is why the equations used
by the boffins at the banks are of limited value. They may incorporate
feedback, but exactly when the positive feedback of greed will become
the negative feedback of fear is impossible to predict. Economic crashes
are foreseen by few; correctly timed by even fewer. But knowledge may
reduce their severity for mathematics gives understanding of feedback.
Mathematics also provides insights into the structures of Nature, and
notably did so in the head of Benoit Mandelbrot.

Mathematics and Structures

> ...the curious repetition of the same forms, of the same design
> almost, in the shape of the falling water. It gave me a sense of
> how completely what seems to us the wildest liberty of nature is
> restrained by governing laws.
>
> – Oscar Wilde, after visiting Niagara Falls in 1882

Benoit Mandelbrot's story[3,4] is a perfect example of the benefits that can flow from allowing selected imaginative thinkers – if only they can first be recognised – free rein. He wrote a paper entitled "How Long is the Coastline of Britain?" Gleick[3] relates how Mandelbrot found two common responses to his question – basically: "I don't know" or "I'll look it up". These responses indicate a lack of imagination for Mandelbrot pointed out that the answer depends on the length of your ruler.

If you walk the whole periphery, you'll get an answer rather close to the conventional value. But if you could roll an atom in and out of every nook and cranny in the coastline, you would get an enormously greater answer. This point may seem patently obvious, but it is in fact contrary to the whole spirit of Euclidian geometry. In the Euclidian philosophy, the shortest distance between two points is a straight line. But taking Mandelbrot's approach, the length of the line depends on how closely you examine it. A straight line might be classically one-dimensional, but if it really had neither depth nor width, it would not be there.

Through nooks and crannies, the area of a surface is increased. By the same token, an increased surface area can be created in a given space through branching (eg 1 splits into 2, 2 into 4). Branching within the lung facilitates oxygen absorption into the vascular system; branching within the vascular system assists the delivery of molecules to and from cells. From Figure 10-2, it is evident that where a quantity is determined by two variables that interact with feedback, the quantity can split into 2 then into 4 and then into 8 values, etc. In contrast to Figure 10-2, in the cases of lungs and vascular systems, the splitting occurs as a function of distance rather than of time. But mathematics does not care whether the variable is distance or time – it is telling us that *simple* rules can lead to these spontaneously branching systems. The genetic programming of splitting is not necessarily as complex as we might assume.

But the rules of mathematics teach us much more. For example, in the reproduction of honey bees, the males derive from the *unfertilised* eggs of the queen – their sole parent. In contrast, female honey bees – workers and queens – derive from the *fertilised* eggs of a queen. Hence, all queens have two parents – their "queen-mother" and the male that fertilised her. As a consequence, if we look at the family tree of a given male, we find that it had 1 parent, 2 grandparents, 3 great-grandparents and 5 great-great-grandparents (Figure 10-3).[5]

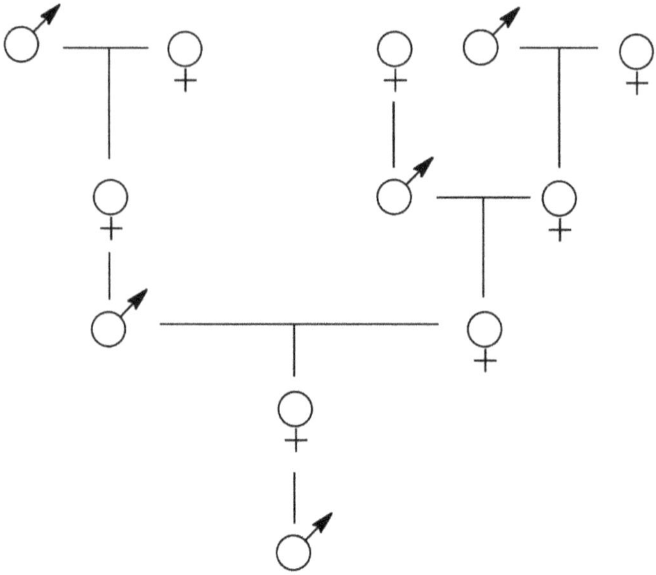

Figure 10-3. The family tree of male honey bees [queens ($♀$) and males ($♂$)]. The great-great grandchild male is at the bottom of the tree; its single parent (queen) at the second level, its 2 (queen and male) grandparents at the third level, its 3 great-grandparents at the fourth level, and its 5 great-great-grandparents at the fifth level.

If the exercise is continued, then the number of ancestors (more generally, "structures") as we pass from the current structure through the generations is:

1, 1, 2, 3, 5, 8, 13, 21, 34, 55, 89, 144, 233, 377, 610, 987…

Two points are illustrated. First, even though sex is only required every other generation to produce the male bee, there is still an astonishingly diverse input of variation on which selection can work to produce fit drones; for fifteen generations ago it had 987 ancestors. Second, in travelling from left to right along the ancestor series, the number to the right is the sum of the two immediately to its left. This is the Fibonacci series – discovered by Leonardo Fibonacci, *ca* 1200 AD.[6] It is extraordinarily relevant to the structures of organisms: lilies and iris have 3 petals; buttercups have 5; some delphiniums have 8; corn marigolds have 13; some asters have 21, whereas daisies are found with 34, 55 or even 89 petals.[5] Genetics takes advantage of simple mathematical rules and, more broadly, much about evolution can be understood in terms of relatively few universal principles.

Societies and the Ratchet Effect

As evident from earlier parts of this chapter, rapid rates of change may bring benefits, but also bring risks. For if a quantity repeatedly doubles, the changes can be difficult not only to accommodate but also their gargantuan nature difficult to intuit. If a large sheet of paper is folded fifty times then what is the approximate thickness of the resulting multiple layers of paper – a metre, 10 metres or a mind-boggling 100 metres?

The answer is easily calculated: take the thickness of the paper as 0.01 cm. The thickness of the multiple folds is therefore $0.01 \times (2)^{50}$, ie approximately 1×10^{13}cm or 100 million kilometres – over half the distance from the Earth to the Sun. So, the "mind-boggling" estimate is too small by a factor of one thousand million. So, repeated doublings of populations soon lead to astronomical numbers, which is why plagues – be they of viral or bacterial diseases, or rabbits or locusts – come; and also go, for their hosts or food supplies cannot be annihilated.

But not all large populations derived through repeated doublings are unstable, for the large reproduction factor may be quenched at some point through internal control. Some of Nature's best survivors are constituted from large end-populations: exemplified by ourselves – a swarm of around 100 trillion cells produced within nine months from the fusion of just two cells. Although less popular as an illustration of the power of the swarm than the ants, locusts, birds, fish and flies of television fame, this example is closer to home – and more impressive.

In human societies, fashions often have startling doubling rates. At the beginning of the twenty-first century, the public wallows in "brands", for images of grandeur are not only now affordable to more, but also highly transmissible. Film and television industries provide them – along with packages of gore and violence – to enhance audience ratings, and hence their profitability and survival chances. There is positive feedback in the loop "film and television hype <–> public love of hype". Through positive feedback, as societies change in the short-term, rise may promote further rise, or rot promote further rot.

People tend to choose what they have been programmed to want; which is great if it is a societal good, but a very slippery slope at the other end of the spectrum. The ultimately degrading use of political control was a spectacle to feed the fantasies of the masses – found in the Coliseum of ancient Rome. There, the mass killings performed and blood lust expressed, from about AD 80 for a period of about four centuries, grew ever more extreme. For maintaining the interest of the populace required that the gruesome spectacles be ratcheted up further and further. Vestiges of these times still entertain humanity: the ritual killing of the bullfight; gladiatorial combat in the boxing ring – fights between professionals often causing serious injury and, occasionally, even death. Primitive facets of the brain change slowly. But the rot stemming from crass idolatry, or a love of violence and gore, often leads to meta-stable states, and negative feedback eventually sets in.

Negative feedback against mad excesses may be stifled in the short-term – especially when violence is a tool of those in power. It is often asked why, in the period 1939-45, more Germans did not spontaneously rise against Nazi tyranny. The answer is clear. As the brutality of controlling regimes increases, the sticking of opposing heads above the parapet becomes uncommon. Propaganda following hard times seduced many Germans and Austrians. The lands that produced Mozart, Beethoven, Schubert, Schumann, Brahms, Mahler and Richard Strauss fell under the control of a barbaric clique. Once Hitler was in a position of power, which he gained through an election held under democratic rules,[7] it was too late – unless you were willing to risk all. The teeth of the ratchet were firmly gripped – for a while.

Feedback on a Grand Scale

Evolution is a tightly coupled dance, with life and the material environment as partners. From the dance emerges the entity Gaia.

– James Lovelock

In 'Gaia – A New Look at Life on Earth', James Lovelock considered all life on Earth, and its interactions with the Earth's dynamic surface, as a super-organism.[8,9] Some organisms modify the Earth by only miniscule degrees, as when beavers build dams. Others modify it in ways that have global consequences.

Lovelock's key point is that there is strong mutual feedback (organisms <–> environment) in what comes to be, impressively illustrated by the biogenic generation of oxygen. Evolution had produced a newcomer: a molecule so toxic that it initially threatened the existence of life; yet a molecule later exploited to produce animals (chapter 4). Thus, the composition of the atmosphere – 79% nitrogen, 21% oxygen, 0.03% carbon dioxide – is a direct consequence of life on Earth. Blood and lungs, and all the organisms in which these structures are found, owe their existence to the oxygen-producing plants. Organisms able to carry out photosynthesis changed the Earth and its contents so much that cities were eventually built.

160

But why is the percentage of oxygen 21% and not, say, 50%? Fires spread more rapidly with increasing levels of oxygen in the atmosphere. There is evidence that about 300 million years ago, the levels of oxygen in the atmosphere reached 35%. At this high level – caused by a great abundance of plants just prior to the evolution of a large population of big land vertebrates – there would be plenty of fires; the Earth's vegetation reduced by the resulting conflagrations. Thus, there is negative feedback control on the viable upper limit of oxygen.

So, if the population of oxygen producers – green plants – becomes very large, they ensure their own partial demise. It is the balance between the production of ca 100,000 million tons of oxygen per year by photosynthesis, and the consumption of ca 100,000 million tons of oxygen per year by oxygen-breathing organisms, that currently maintains the level of oxygen in the atmosphere at 21% – feedback control on a grand scale. Exploitation of resources commonly leads initially to excess, later curbed in the service of survival. Exploitation through competition then curbed through feedback is a recurring theme that influences the populations of organisms and their behaviour. The behaviours of organisms that emerge as survival strategies are explored next.

11

Behaviour

Selection and Survival

> 'The medical techniques used in the Vietnam War are now being
> used in civilian life,' said Dr Garen Wintemute, an assistant
> professor of medicine at the University of California at Davis
> and former medical director of a refugee camp in Cambodia.
> 'There's no difference. And that wasn't the case until the advent
> of assault rifles…
>
> 'His comparison to Vietnam was echoed in interviews
> with more than a dozen other doctors and paramedics across the
> country, who described exploded organs and pulverised bones,
> the flood of internal bleeding and bodies riddled from high-
> velocity, rapid-fire assault rifles…
>
> 'The military-style assault rifles do more damage per bullet,
> and fire more bullets before they must be reloaded, than do the
> pistols that once prevailed among gangs in the inner cities. With
> a muzzle velocity of more than 2,500 feet a second, as against
> perhaps 800 feet for a pistol, the bullets are designed to tumble
> upon impact, shredding bones, organs and vessels in their path…'
>
> – 'New York Times', 21 February 1989

The *New York Times* report underlines the long journey from primitive
bacteria to primitive humans. Selection has been uncompromising in
wiping out antecedents and weak variants – reflected within limestone
cliffs; in the coal and oil made from fossilised organisms; in the extent of
the Barrier Reef, visible from space and made – with the exception of the
living coral near its surface – from dead stock.

But the gee whiz observations grossly underestimate what has been
eliminated through selection. A turnover of an average of a trillion tons of
biomass per year over the last billion years is a very conservative estimate.
Therefore, the amazing variety of organisms that survive today should be

looked in the light of much more than 1,000,000,000,000,000,000,000 tons of biological precursors – materials in the surface of the Earth turned over again and again. So the extant machines of Nature are awesome in performance; inevitably exhibit behaviours that both exhilarate and disgust. It is because enormous numbers of parental ancestors were survivors on a competitive Earth, while countless others were being discarded, that there exist the awesome biological machines of today.

An understanding of the behaviour of organisms – their survival strategies – requires an understanding of the processes of selection. In these processes, individual genes are not commonly targeted. Rather, the targets are individuals and groups of individuals. As Sydney Brenner once put it to me with his usual clarity: "If you are being chased by a man with a hatchet, your survival is dependent on how fast you run." Fleetness of foot is determined by very large numbers of genes, and also by the environment in which they are expressed. Individuals, and groups of individuals, are commonly targets of selection – as in the above *New York Times* report.

But individuals are not selected as entities that can be defined in the longer term because they die. To the best approximation, the definable unit that is selected is the "gene" – a chunk of DNA sufficiently small to be inherited through many generations.[1] So, through all this targeting, it is the genes – the repositories of information that are copied – that are selected for representation in the individuals of the future.

Efficient function that depends upon the expression of large combinations of genes is selected through the targeting of organisms. In the wild, there is a predator/prey relationship between the browsers and the carnivores. Thus, there has been great pressure in lions for the selection of the combination of genes that promote stealth of approach to prey, speed to capture, forward-pointing eyes, power, and other aspects of efficient killing.

In gazelles, the behaviours that counter these abilities are co-selected: speed to escape, eyes that face to the side for a large field of vision, and sensitive hearing. Thus, the form of gazelles is determined to an important extent by the form of lions, and vice versa – just as the forms of flowers and of the pollinating insects are mutually dictated. Through their mutual dependence, each form imposes an influence on the other which pushes each of them to new heights of performance.

In mass extinctions, it may seem that "survival of the fittest" becomes irrelevant; the forces unleashed so powerful that few survive. There have been at least five major extinctions in which large numbers of species met their demise. The greatest – the Permian, 250 million years ago – is estimated to have led to the demise of around 90% of the species existing at that time. About 65 million years ago, the impact of a large meteorite – perhaps coincidental with contamination of the atmosphere by high volcanic activity – led to the extinction of the large dinosaurs. The combinations of genes that produced large dinosaurs have not re-evolved, although a similar package did survive and led to the birds of today.

However, even in mass extinctions, some do survive; fitness still relevant. Some bacteria live deep within rocks – wonderfully adapted to survive the evolutionary pressures associated with catastrophes at the exposed surface of the Earth. Genes for the formation and function of such single-cell organisms survive at the expense of those coding for big and exposed organisms.

The battle for survival is something to which all are subject. Therefore, an analysis of the behaviour of other species should further an understanding of many things done by humans. So, next, there is consideration of the way in which genes and environments together determine organisms; a look taken at a few of the derived behaviours. In all cases, strategies that promoted survival – some beautiful, others brutal – were selected, with no holds barred.[2]

Determinants of Form and Behaviour

That each effect must have a cause allowed an understanding of what happens in the Universe. Cause and effect are accepted in the operation of inanimate systems. Cars coming out of the same factory production line at the same time have the same form – except insofar as there was minor variation on the line. The behaviour of the car on the highway is then determined by the totality of the inputs: what went into the making of the car, and the environmental influences that have worked upon it from birth and are currently at work – primarily the driver.

The form and behaviour of an organism, including a human, must be determined in the same manner: its form at birth determined by genes and the environment in which they were expressed; its form and behaviour as it passes through life, both by its genes and the environmental influences that have worked upon it from birth. Both genes and environment determine, and each acts upon the other. Although there are hazards in attempts to unravel this convolution, useful guides can be derived.

The production lines for an organism are large in number, being successively switched on, one by the other. But, given standard environments for development, form at birth – spider, snake, chimpanzee, human – is genetically determined. Yet the genes expressed at birth do not represent the whole of the library. Some sentences are read only later as the organism passes through life, to dictate the processes of further growth, maturation and decay. In humans, the genes for sexual development are expressed ten to seventeen years after birth. Internal clocks trigger new bulges after long time spans. But genes determine not only form but also many aspects of behaviour.

Genes and Behaviour

Spiders and snakes, whether reared in the wild or in isolation within a box, both behave in ways highly characteristic of the species. The same is true, but to a lesser extent, of chimpanzees and humans. If cruelly restricted in space and companionship, the behaviour of each of these species is still either chimp-like or human-like. Some behaviour is gene-dominated – a conclusion reinforced by experiments in which chimpanzees are reared in human environments. They still behave in chimp-like ways, despite the adoption of a few characteristics from human culture. Since humans and chimpanzees share about 97% of equivalent genes, relatively small differences in genes can clearly affect behaviour a great deal.

But behaviour is complicated by the variable extents to which environmental influences are stored; and by the way in which the information in the store is then processed before use. It becomes increasingly adaptable as the extent to which information is stored and processed is increased. Although useful, the pigeon-hole change of hardwired –> adaptive involves very fuzzy boundaries; in reality, an evolutionary continuum.

Behaviour – Hardwired and Adaptive

> *When I was 14, I regarded my father as so ignorant I was embarrassed to have him around. By the time I was 21, I was amazed how much he had learned.*
>
> – Mark Twain

If an organism has little or no facility for memory and its filtered use then its behaviour is said to be hardwired. Very primitive hardwired behaviour occurs when parts of a plant move towards sunlight, or a bacterium swims towards food. The signals of sunlight or food always produce a specific response when acting upon specific devices.

Even where facilities for memory and its filtered use are advanced, some remarkably complex behavioural traits are still hardwired – presumably because they are ancient in origin and/or crucial for survival. Hornbills provide a striking example. Parents-to-be build a nest in a hole in a tree and then largely fill in the hole, with the female inside, the male outside. The residual hole is just large enough for the male to feed the female and the hatched chicks within the nest. When the female has finished her parental care, she enlarges the hole to regain her freedom. But, at this point, the chicks once more immure themselves – the risk of predation by monkeys and snakes so reduced.

So that fully developed chicks can escape the nest, their behaviour is reversed – material now pecked out of the hole. The opposing behaviours of "hole-filling" and "hole-enlargement" are precisely programmed as a function of the age of the chick. Thus, chicks that are produced from the later hatching eggs perform "hole-filling behaviour" at the same time as chicks from earlier hatching eggs perform "hole-enlarging behaviour". The lack of logic in these opposing activities emphasises their truly hardwired nature.

DNA tells spiders how to construct webs; instructs a dog to roll on its back upon small dead animals or products derived from them. The evolution of the latter behaviour probably lies in disguising the scent of the dog by one less alarming to potential prey. Dogs domesticated for a hundred generations retained this genetically programmed trait. A kitten removed from the mother at an early age, and subsequently living with humans and isolated from other cats, does not need to be taught the characteristic hunting behaviour with the low, stealthy approach; nor the grooming behaviour with characteristic movement of limbs and use of the tongue. All programmed in the DNA.

Our sense of awe increases when we consider that the DNA not only tells an organism to do these things but also – given typical environments – when to do them. The queen wasp has a DNA program that not only tells it how to fly but also how and when to search for a suitable nesting place. A bird does not need to be taught how to build its nest – the molecules of its DNA contain a program that tells it what kind of materials to collect, how to assemble them into an appropriate structure, and when to do these things. A child walks at around one year, starts to talk – given that it hears others speak – around two years, and sexual behaviour is switched on in the early teens. That so much behavioural detail can be, when given the appropriate environmental prompts, stored directly in DNA sequence must surely astonish.

Human behaviour is more influenced by standard response patterns than commonly accepted. Konrad Lorenz pointed out in 1943 that parental instinctive responses are induced in adult humans by baby-like features found in a wide variety of birds and mammals. Descriptions of "cuddly" and "cute" are applied to animals that have a short face in relation to a large forehead, protruding cheeks and large eyes. These features, found in human babies, render even a large puppy more likely to be cuddled than a small adult dog. Walt Disney used, and benefited from, the response. Even where the environment is rather unusual, the hardwired influences of genes are still evident. For, as again shown by Konrad Lorenz, young animals can not only be induced to follow their parents after receiving auditory and visual stimuli from them but this response is sufficiently automated that Lorenz could induce young geese to follow him as an "adopted parent".

Genes show their power when land mammals immediately struggle to their feet at birth, and when human babies pull themselves upright with the help of supporting objects a few months after birth. But, even here, they are receiving environmental prompts: to a degree, adaptive behaviour because the environment is tuning the genetic guidance.[3]

Some behaviour conventionally regarded as hardwired is in fact quite adaptable – as for some activities of the honey bee.[2] Honey bees have a range of flight, navigational and communications systems that make an aircraft look rather pathetic. In terms of the time taken by the bee to fly its own length, it flies faster than a plane; and has the ability to land on surfaces vertical, horizontal and complex. Its automatic and programmable navigational systems allow it to locate and avoid objects in space, and to perform approaches to these objects by amazing control of its velocity in three dimensions. Its sting can put organisms of many millions times greater size to flight. Military top brass, eat your hearts out.

The automatic construction of this biological machine is programmed in a fertilised egg around 1mm in length. The machine can locate pollen and honey sources at distances 100 thousand times its length from the hive. It then not only successfully relocates to the hive but also informs the other bees in the hive of the location of the food source. The bee also remembers on the following day the field that was productively foraged for honey on the previous day.

The German biologist Karl von Frisch decoded some aspects of these amazing feats in 1945. The worker bee, on return to the hive, performs a repeated dance, approximating to a figure of eight. When she traverses the almost straight part of the "8" that connects the two loops, she wiggles her body and also vibrates her wings to emit an audible buzz. Where the dance is performed on the landing board of the hive, the straight run points towards the food source. Bees perceiving this information retain it as referenced to the current direction of the Sun, thus allowing them to remain on course during flight. In contrast, if the dance is carried out on the vertical combs inside the hive then the angle between the vertical of gravity and the straight part of the run now convey the angle between the Sun and the food source.

The time taken for the straight part of the run communicates the distance of the food source. In one species of honey bee, a straight run lasting one second indicates a range of about 500 metres, and one lasting two seconds indicates a range of about 2 kilometres. Additionally, foraging information is transferred when the buzzes emitted by one bee are detected and interpreted by a second – processes analogous to speaking and hearing. The signalling is sufficiently effective for surrounding bees to leave the hive within minutes, and then to search for food within 20% of the correct distance.

Thus, bees are able to receive, store and interpret, information on direction and distance. Insofar as bees learn from their colleagues and adapt their subsequent foraging behaviour, they are intelligent creatures. For one dictionary definition of intelligence is "the ability to adapt to new situations, and to learn from experience". Bees make choices, but clearly are not "free". As to whether a bee is or is not "conscious" depends on the dictionary definition used. One dictionary definition of "conscious" is "internally known". The location of the food is internally known for otherwise the bees could not repeatedly find it. So, by this definition, a bee is conscious of the location of the food.

So, an apparently largely "hardwired" bee does many remarkable things; but it is not exceptional, for survival strategies are typically amazingly sophisticated.[4] In some cases, a mathematical approach can show that problems are solved by animals through solutions – however one may wish to classify them in the gradation hardwired –> adaptive – that are indeed the best ones. For example, starlings may lack the ability to manipulate numbers but get the right number in a foraging strategy.

In delivering grubs – mealworms – to nestlings,[5] starlings do not carry grubs singly for they are able to carry several. But what is the number of grubs for the optimum collection/carrying strategy? The beak is opened to part vegetation and so uncover grubs. But this strategy becomes more difficult with increasing numbers of grubs in the beak. Experiments show that the searching time for a given number of grubs increases rather rapidly once large numbers are held in the beak; as in a supermarket without a large basket, so many purchases that you drop as many as you pick up.

The optimum collection/carrying strategy must also depend on the distance back to the nest. For, only one loaf of bread per visit to the supermarket might be bought if it were on the doorstep, but several might be bought if the supermarket were several miles away. For a fixed travelling time, a starling has to optimise its load of grubs so that the number of grubs delivered to the nest in a given time (say measured in grubs per minute) is a maximum. The equations used to provide the answer are based on a classical margin value theorem, known from economics. Does the starling get the right answer? And, if so, how?

In 1984, Alex Kacelnik tested the model. He found that not only did the starlings carry larger loads when collecting grubs at greater distances from the nest but also there was an impressive quantitative correspondence between the size of the load and that predicted by the model.[5] How do the starlings "know" the optimum load? They are programmed by their molecular systems to work in the most efficient way. The efficiency does not derive from deduction; it derives from variation, fitness and selection.

The hypothesis of the survival of the fittest tells us that inefficiency relative to a competitor will not be tolerated. In the long run, the benefit of the behaviour must be greater than its costs, as shown by Nikolaas Tinbergen in a study of gulls. When chicks hatch, broken eggshells in the vicinity of the nest provide evidence to a potential predator of the relatively defenceless meal. Nikolaas observed that, after about an hour, parents in a colony of black-headed gulls carried the eggshells away from the nest. Hypothesise that this behaviour has evolved because the chicks are more likely to survive, and the parent thereby more likely to pass on its genes. Being a first-class scientist, Tinbergen[6] tested a necessary consequence of the hypothesis. He showed that predatory crows were indeed more likely to discover, and eat, eggs in the colony if broken shells were left lying near the eggs.

But why does the black-headed gull wait for about an hour before it removes the eggshells? Would it not be more efficient to remove the shells immediately after hatching? The potential cost of the latter strategy is not difficult to envisage – a predator may step in while the parent is away dumping the eggshells. Indeed, Nikolaas discovered that newly hatched chicks, still slimy after birth, provide a convenient meal for cannibalistic gull neighbours. An hour or so later, when dry, they are much more difficult to swallow. So, the mother black-headed gull waits for about an hour. The behaviour might be considered clever; conditioned is more appropriate – as when beavers build dams.[7]

Large and extremely complicated collections of neurons allow awesome storage and processing of information from the environment;

behaviour – as for humans – is then more flexible. But the degree of flexibility in the behaviour is determined by the totality of its past and present inputs; flexibility likely only to the degree that survival has been served. Significantly different forms, and markedly different behaviours, are evident in genetically identical human twins – derived from the division of one fertilised egg into two embryos. Even when the twins are within the mother, there are finite differences in the environments of the two developing systems. The twins may have strikingly similar faces but they are not identical. Following birth, even a single severe chastisement of a child can precipitate a lack of confidence that is then carried forward and subsequently greatly amplified. Through positive feedback upon small beginnings, reticence *vs* confidence may be born. The twins have different behaviours.

It does not make sense to attempt even semi-quantification of nature *vs* nurture arguments in terms of "intelligence is 50% genetic and 50% education"; nor "musical ability is around 25% genes and 75% environment". The semi-quantification is as nonsensical as "the sound of a violin is 5% due to the bow and 95% due to the violin". These outcomes depend not simply on the isolated systems (genes and environment, or violin and bow), but also on the interactions between the pairs. Had Mozart never had access to a musical instrument, nor heard a tune, then his oft-cited "born musical genius" would have been zero. His genes could only give potential.

Crows – like humans – show behaviour that is remarkably hardwired in some tasks, but remarkably plastic in others,[8] some of which must have evolved very recently. Carrion crows living in the environs of towns use passing vehicles to crack tough nuts. They do this at traffic light crossings, waiting patiently with human pedestrians for a red light before retrieving their reward. New Caledonian crows make at least two types of tool to aid them in their search for food. They make probes to extract invertebrates from leaf detritus by sculpting leaves with their beaks. Twigs are used to make hooks which are then used to poke grubs from holes in trees. In captivity, a New Caledonian crow, unable to gain access to food at the bottom of a cylinder that the crow was too big to enter, learned how to bend a piece of straight wire into a hook.

These studies of crows suggest that they understand their physical and social worlds in a similar way as "understanding" is used in describing the behaviour of monkeys, apes and humans – albeit all to different degrees. Although the last common ancestor of crows and primates lived about 300 million years ago (before the evolution of complex cognitive abilities), an individual crow can come up with solutions to problems in a way that establishes that they, like humans, have novel ideas. So, complex cognitive abilities have evolved independently in distantly related species with very different brain structures in order to solve similar problems. As with humans, there is the combination of previously available bits of information to give innovation. Original thinking is not the domain solely of humans.

Social Behaviour

> *What do I think of Western civilisation? I think it would be a very good idea.*
>
> – Mohandas Gandhi

Whenever there is the evolution of new aggregates – of molecules, cellular components, cells, individuals or nations, to name but a few – new properties emerge. The aggregates are then subsequently commonly selected with increasing cooperation between the units that aggregate. The stability and population of the aggregated form is thereby enhanced. This trend provides a basis for the evolution of social behaviour. And be it in bees, black-headed gulls, beavers or humans, sociality imposes additional behavioural constraints.

Sociality began in the inanimate world. In the process of crystallisation, impurities – molecules that do not conform to efficient packing in the crystal lattice – are preferentially excluded. Within the crystal, there is discrimination against non-conformist types. In the living world, cooperation begins also at the level of molecules – evident in the cooperatively folded structures of proteins. The shapes of the 20 amino acids that form the proteins have been selected such that they can normally fit together to give a well-packed and highly cooperative protein structure. Even more startling cooperation appears to have evolved between enzymes and the molecules they transform.[9,10] Not trivial for, without this cooperation, food could not be processed and animals could not function. The molecules of biology are commonly social.

At a higher level of organisation, cooperation occurs between components of the cell. Further still up the hierarchy, it occurs between different cell types within animals. A human has a good chance of survival when its various types of cells cooperate with each other – each performs its function and remains within defined areas of the complex organism. But once breeding is finished, such cooperation serves Nature less well, for the organism has served its key function. Cancer is the disease in which a given cell type divides out of control. Cooperation between some of the differentiated cells has been terminated and replaced by competition. The rogue cancer cells compete in niches that they would not normally occupy, as when breast cancer leads to "secondaries" in bone.

At levels of even greater complexity, cooperation occurs between the individuals of social groups;[11] but the cooperative constraints acting upon the individual still sit uneasily with the benefits that can derive through selfishness. A social group may need military conscription to promote its survival. Patriotism is a cooperative trait pressed by environmental influences, and presumably also promoted by the genetics of social behaviour. But the individual often perceived the true hazard of war – hence white feathers and the firing squad.

Cooperation within extended groups of humans is a relatively recent phenomenon. But, throughout history, the cooperation required of individuals to enhance survival within social groups inevitably discriminated against individuals whose behaviour could not be moulded to conformity. In the longer term, through this discrimination, increased cooperation was genetically selected. The benefits of cooperation within a *social group* became so important that adverse genetic selection occurred for some of its *individuals*. The genetic constitutions of the Bourbon and Hapsburg families suffered through the inbreeding that the social group forced upon them, for cohesion was enabled through heads of state with "pure royal blood". Genetic deformities and poor mental capacities appeared. Even allegedly powerful individuals are manipulated by the overarching needs of social groups. The queen of the termite colony is not alone in her eternal egg-laying labours.

We have arrived at the core of the biological basis of social behaviour. The doyen in this field – that of socio-biology – is Edward O Wilson.[2] He found that animal behaviour is strikingly consistent with well-optimised survival strategies. Guppies enhance their survival chances through aggregation into large shoals, believed to benefit foraging and perhaps also as a defence against predators. Individuals are served through this cooperation, but are still brutally competitive when circumstances demand.[12] A single pregnant female guppy was introduced into one aquarium, and fifty mixed individuals into a second that provided the same environment as the first. Both populations converged to nine individuals and stabilised there, a greater population density being avoided through cannibalism – behaviour with feedback.

Aspects of cooperative and benign behaviour are expressed at relatively low populations, and survival thereby enhanced. High population densities are reduced by brutal competition – back towards the size of the entity that is the most stable under the given conditions. Social groups can get too big for their boots, and downsizing follows. Humans with increasingly dense populations, once more take note.

Cannibalism is common among the social insects. It is used among termites not only to preserve nutrients but also to regulate colony size. The colony wins. Among humans, it was known in New Guinea where it may have provided a source of protein in an otherwise protein-deficient diet.[13] Perhaps here also, the colony won. It occurred in the twentieth century among small groups of males facing starvation in extreme environments. Here some individuals won at the expense of others. Perhaps not surprisingly, Edward Wilson was not popular with everybody. Some thought the evidence he assembled demeaned humanity. So much so that in 1978 a pitcher of iced water was poured over his head as he stood up to give a talk. But his findings were evidence-based and have largely stood the test of time.

Up to the middle of the twentieth century, social groups made limited contact with each other and competition was the dominant force in selection between them. There then followed enormously greater contacts between the Earth's social groups – through science and technology, communications, trade and migration became almost global in their extents. As for the molecules in the cell, evolution eventually hones each sub-unit to perform its dedicated task within the whole so as to promote survival. But, as the honing process proceeds, there is a delicate balance between many possible outcomes; cooperation reduced when others are perceived to behave selfishly – hence a route to wars.

War Games

Unrestrained competition within a social group risks mayhem. The reason why mayhem is the exception rather than the norm may be understood from points made by John Maynard Smith,[14] based on ideas derived in large measure from Bill Hamilton and Robert MacArthur.

The competition within the group may be for food, living space, a mate for reproduction and so on. An individual gaining the best of these commodities is likely to improve its chances of passing on its genes to the next generation. Power may in the short-term enhance access to such commodities – hence the drive to be top dog/politician/CEO. But when **A** encounters **B**, why does **A** not simply exterminate **B**, thus ensuring control of goodies that otherwise would have been shared between them? To be sure, murder exists in human – and some other – social groups, but it is the exception rather than the rule. A little reflection suggests why this is so – if two individuals behave in the ultimately hawkish manner towards each other, one of them is likely to be wiped out. Put in another way, if hawk meets hawk, there will soon be blood on the carpet – and that is a poor survival strategy. So what is a better way to behave?

It might be considered that it would be better always to behave in a dove-like manner. For, if you defer to a hawk, then at least you might avoid serious injury. But this is not so, as can be seen by considering an initial population consisting of only doves. The next generation will be affected by mutations, crossovers between parental DNAs, and variations in gene expression as a consequence of different environments. Thus, it may contain one or more individuals who are more aggressive than the population at large. Among the sea of doves, these individuals win all that is going, breed successfully, and many hawkish individuals are likely to be produced in the generations that immediately follow. The number of doves is drastically reduced, illustrating the inadequacy of totally dove-like behaviour.

The best policy might be not to join either a hawk or a dove group, but rather to modify behaviour according to the circumstances. Indeed, a model in which most individuals are "conditional strategists" better approximates the kind of behaviour seen in social groups of mammals. A conditional strategist is an individual whose behaviour depends on the behaviour of the second party, but commences all encounters by adopting a dove-like strategy. If it turns out that the second party adopts a hawk strategy then the conditional strategist can behave likewise. Similarly, if it turns out he/she has met a dove, he/she can then behave like a dove.

Striking aspects of the conditional strategist model are seen in human encounters. For example, in the formal and polite behaviour usually displayed at a first meeting, or even at the outset of meetings that are not for the first time. If, after some time, one party makes an aggressive move or statement, the second party assesses whether it is worthwhile to respond in kind or whether to acquiesce – usually taken to imply the operation of free choice. But the same behaviour is seen in rutting stags, in the fighting between male walruses – enough to determine access to females but not too frequently leading to death. The theory of war games tells why we not only cooperate as well as fight, but also how we are programmed – as for rutting stags and walruses – to adopt a strategy that is good for survival.

Within social groups, burgeoning interdependence promotes cooperation, for it is not a good strategy to wipe out those on whom your survival in part depends. Large herds of gnus reinforce the message.

Gnus Outnumber Lions

The promotion of cooperation within groups of organisms as a natural consequence of Darwin's ideas is neglected. It is often argued that the Nazis took heart from Darwinism because of its common expression as "survival of the fittest". They seem to have believed that this meant that strong aggressors, and physically fit Aryans, had the best chances of survival. If this is so then they got it completely wrong – as they did with almost everything else. Their own demise cogently makes the point.

In fact, cooperation in social groups is a very successful survival strategy. Predators do not preclude success among gazelles and gnus, for these gentle browsers greatly outnumber lions and leopards. These observations are entirely opposite to the common perception of Darwinism. Nature is not simply read in tooth and claw. Survival of the fittest also involves cooperation within social groups; involves gentle care in animals that are closely genetically related. Even those who regard themselves as sophisticated Darwinians often fall into the trap of expressing the philosophy in terms of "the weakest go to the wall"; erroneous, for natural selection has currently brought the gene package of the fearsomely strong tiger close to extinction.

Thus, to describe attempts to kill, or grossly discriminate against, those who are weak as "social Darwinism" is an obscenity that should vanish from the modern vocabulary. For it is the principles enunciated by Charles Darwin that allow us to understand why gnus can outnumber lions; why leopards can be fewer in number than their prey; why violent dictators pursuing a policy of "racial cleansing" may meet an early fate. It is a judicious combination of both cooperation and competition that provides the best survival strategy within social groups. Darwinism is sometimes the survival of the gentlest.

Knowing Tomorrow

A superficial analysis suggests that we can only know the past and never tomorrow. But tomorrow and today follow the same basic rules, and survival is aided by this knowledge. This fact encouraged Nicky Clayton, an animal behaviourist at the University of Cambridge, to use the provocative lecture title *Do Animals Have Memories of Tomorrow?* She observed that scrub jays that had seen food being stolen from hidden caches would change the location of their caches in an attempt to avoid future thefts; those that had not observed the thieving behaviour did not adopt this strategy;[15] quite a cognitive ability within a group of animals commonly patronised as birdbrained. The jays are able to memorise experiences from their recent past to provide survival advantages tomorrow.

Like the scrub jays, we humans "know tomorrow" and, with our much bigger brains, devise fiendishly complicated strategies in attempts to benefit. In the twenty-first century we know that rich polluters of the environment do not have a leg to stand on if faced with a greater power. The disaster of the BP oil spill into the Gulf of Mexico in 2010 carries all the trademarks of groups of primates exercising vested interests – behaviours reflecting both a "roll over" to a greater power and the exercise of that greater power. Companies frequently economise on safety since survival at the expense of competitors may be served. BP was no exception – there was much evidence that they took short cuts on safety procedures. The immediate disaster led to the loss of eleven lives. BP initially spoke of a leak of circa 5,000 barrels a day. It was later evident that, although still uncertain, it was closer to 50,000 barrels a day.

As the immediate environmental impact of the leak became clear, the ire of US citizens increased, and the White House took up cudgels. US claims against BP were estimated as $20-35 billion – an enormous sum. For example, if there were one business qualifying for compensation every 100 yards of the whole US Gulf coastline (a total of about 30,000 businesses) then $30 billion would allow a compensation of $1 million per business. These numbers represent power – not least of the lawyers who swiftly set up hotlines for custom. The comment of Tony Hayward (CEO of BP) that the spill was "a drop in the ocean" was politically incorrect and insensitive and resulted in his vilification. Yet it was scientifically correct, for the volume of the Gulf seawater is roughly three billion times that of the oil spill. Add to this the fact that components of crude oil are natural substances which many organisms have used as a food source from time immemorial and, no surprise to the scientifically trained mind, the oil spill was gobbled up much faster than the doomsters had predicted.

The complexities of social systems are ignored in attempts to blame, to punish, to benefit and to occupy "the moral high ground". First, President Obama wanted to know which "ass to kick" for the errors, suggesting that Tony Hayward should resign. The political pressure was such that he did. But when, during the first half of 2010, US drone attacks killed far more Pakistani and Afghani civilians than militants, President Obama as Commander-in-Chief of US forces did not resign. Second, Kenneth Feinberg, a Washington lawyer, was appointed by the US administration as an "independent administrator" to ensure "impartiality" in the award of compensation from the $20 billion that BP had put in escrow. Feinberg quickly commented, "We have got to err on the side of the claimant" – some impartiality. Third, the 1988 Piper Alpha disaster in the North Sea occurred with the loss of 167 lives. Yet, unlike Hayward, the CEO of the operating company (Occidental Petroleum, with headquarters in Los Angeles) was not called to face grieving relatives in front of television cameras. Fourth, BP was producing oil from the Gulf to satisfy massive US consumption and with the approval of the US government. Politics was partly driving the US response, for the previous (G W Bush) administration had suffered from an under-reaction to the Hurricane Katrina disaster.

The example illustrates how one super-organism (here a powerful country) is able to punish another super-organism (here a powerful company, but a system weaker than the country) according to their relative punching weights. If this conclusion is true then, when powerful establishments err and the weak suffer, the boot must be seen to be on the other foot. And it is. When a plant in Bhopal, India, leaked methyl isocyanate in 1984, the leak caused the deaths of from 2,000 up to 16,000 people (gross uncertainties). The plant was owned and run by Union Carbide India Limited, a subsidiary of the US giant Union Carbide. Safety considerations in the factory were astonishingly lax. The final compensation (including interim relief) for personal injury, for the majority, is reported as a mere Rs 25,000 (US $830); and for a death claim, the average sum paid out was a miniscule (even allowing for differences in living costs) Rs 62,000 (US $2,058).[16]

The above examples refer to the US simply because it is currently the world's most powerful nation. But the principle is essentially universal. In 1972, members of the British Army shot twenty-six civil rights protesters of whom fourteen died, in Derry, Northern Ireland. A first investigation by the British Government largely cleared the soldiers and British authorities of blame; it was subsequently described as "a whitewash". Only in 2010 did a second enquiry find that all of those shot were unarmed, and that the killings were "unjustified and unjustifiable".[17] On the publication of this second report, the British Prime Minister, David Cameron, made a formal apology on behalf of the United Kingdom. But, thirty-eight years after the event, no criminal charges have been made, and no compensation paid.

Together, these examples and hundreds of others involving countless countries that could be cited in a similar vein illustrate how powerful groups serve their vested interests in the battles for survival; how our primate natures are much in evidence. What other animals do does help in understanding the human condition and behaviour. Karl von Frisch (1886-1982), Konrad Lorenz (1903-1989) and Nicolaas Tinbergen (1907-

1988), whose work has figured prominently here, had in common an intense curiosity as to why animals behave in the way that they do. For their great contributions to our understanding of patterns in social behaviour, they were jointly awarded the 1973 Nobel Prize in Physiology and Medicine. The work started by them requires a reconsideration of what we ourselves are – a challenge taken up by Desmond Morris when he portrayed the close relationships of human behaviour to those of our primate relatives.[18,19] Although humans are unique, so are all other species; all honed by common forces; and with behaviours so amazing[20] as to challenge our wish to be set apart.

Here, and later, a given strategy may not to be blamed or praised. Rather, behaviour is simply observed. For since only our past allows us to "know tomorrow", our behaviours – whether loving or disturbing – do not come with the freedom implied in orthodox views. Even when new inferences arise from past knowledge *via* combinatorial play within the brain, the molecules are masters. The supporting evidence comes next.

12

Why Freedom Dies

Humans – Constrained Machines

*All these young souls were passengers in the Durbeyfield ship
– entirely dependent on the judgement of the two Durbeyfield
adults for their pleasures, their necessities, their health, even
their existence. If the heads of the Durbeyfield household chose
to sail into difficulty, disaster, starvation, disease, degradation,
death, thither were these half-dozen little captives under hatches
compelled to sail with them – six helpless creatures, who had
never been asked if they wished for life on any terms, much less
if they wished for it on such hard conditions as were involved in
being of the shiftless house of Durbeyfield. Some people would
like to know whence the poet, whose philosophy is in these days
deemed as profound and trustworthy as his song is breezy and
pure, gets his authority for speaking of 'Nature's holy plan'.*

– Tess of the D'Urbervilles by Thomas Hardy, 1891

The preceding concise description of the journeys of matter, both without
and within organisms, and of the principles that determined its routes,
serves to require a fundamental reconsideration of the human condition.
It takes us from truths that can be tested to those that are also taxing,
for it serves to encourage inspection of humans as biological machines.
From the outside, they may give the impression of being free agents. But,
peering into their interiors upon surgery or autopsy, a more appropriate
description seems to be "awesome machine".

Yet the description raises hackles – a book on truths must say many
things that "cannot be said"- for it "demeans amazing beings". But polar
bears and pandas are also amazing creatures; chimpanzees and humans
strikingly alike because they shared their evolutionary history to an extent
of 99.9%, and share equivalent genes to the extent of about 97%. Can
chimpanzees lack free will while humans possess it; does a larger forebrain

confer free will? Or, is Thomas Hardy correct to infer that the outcomes for his Tess were determined by the totality of her circumstances? Determined by her inheritance from ancestors, and by the manipulation – through her environment from womb to tomb – of her genes (both in the mutations of some, and relative levels of expression, and activities of many)?

Analogies between the behaviour of humans and other mammals are too striking to be ignored: the dominant wolf walks tall; its dominated colleagues grovel at lower levels – sounds like a familiar trait. All creatures benefit or suffer in genetic and environmental lotteries – steeped in games of chance. From the way in which selection has operated in the lotteries, there are strong constraints upon what humans can be. And so the first shibboleth to tumble is that of free will. Voltaire's maxim *"Liberty* of thought is the life of the *soul"* is shown to be false. But it is a better approximation in some circumstances than others; and, as seen later, a maxim that, under no circumstances, will be consigned to the dustbin.

The Falsehood of Free Will

> *Free will, if you pause to reflect,*
> *violates cause and effect.*
> *So, if you choose to pause*
> *and contemplate cause,*
> *the choice isn't yours, I suspect.*
>
> – Colin Beard, Californian scientist
> A limerick dedicated to Dudley in the name of their friendship

All species are moulded by common principles. Given a belief in cause and effect then they are what they are as a consequence of (i) all the states, of which there is genetic memory, that their ancestors passed through and (ii) all the states, of which there is memory, passed through during the course of their lives. In sum, determined by genes and environment – the only means by which information has entered, and can enter, the system. Novel information can leave the system, courtesy of combinatorial play within brains. But the brain did not evolve as an agent of freedom; built by the environmental influences that acted upon ancestors, it evolved to promote successful function.

But, within religious groups, the view "God gave me the freedom to be good or evil" is often expressed. And among all individuals and groups, secular or religious, free will is very widely accepted. It forms a crucial basis upon which society operates:

Barrister: *"Did you do this of your own free will?"*
Accused shifts uncomfortably.

And, in view of the brainwashing we receive from birth on this subject, perfectly understandably. But do we possess it?

The UK Oxford Dictionary definition of free will is "power of directing our own actions without constraint by necessity or fate". But human actions are clearly restrained by necessity and fate for there is no freedom in physiological function. In the apparently free choice by a sated human to sip, or otherwise, from a glass of water, fate provided the involved brain and muscles. Whether they function, or malfunction, it is your fate. And if the human system were in the diametrically opposed state – drastically parched – the necessity to drink is also imposed by fate.

Among the trillions of events that determine the system of a baby at birth, the baby does not choose a single one. The necessary information, and the consequences of its expression, came from the parents. So self-evident is this truth that free will is decreed to arise during subsequent development – on the basis of "freedom of thought". But the nature of thought is, as for automated behaviour, determined by our origins. Evolutionary pressures and selection fashioned the fore-brain. And within this novel structure, new properties evolved – self-awareness and conscious thought. But these properties did not evolve to give freedom; rather the contrary, because to behave without constraint is shown by history to be very dangerous. They were selected to promote survival.

When a baby makes choices in the first week after birth, those choices are completely constrained by its genes and environment, now including the environment into which it is thrust in that first week. The argument can be extrapolated forward, through a large succession of very small time intervals. The conclusion is always the same. Our decisions at any one point in time are a consequence of what went before. At each and every point, we become a system dictated by its past, and respond accordingly to new inputs.[1]

The determination of humans by past inputs extends to the determination of the new designs that they may generate through "intelligence". "Intelligent human design" is the bringing together of many different things and selecting from among their myriad possible combinations – variants – those that provide the best function. Variations and selections that lead to pendulums, coiled springs and numbers; and variations and selections that then put these, and many other devices, in combination, to produce watches and clocks.

"New ideas" are similarly derived – not through free will but rather when the brain brings into combination a number of different concepts.

These concepts are those of past inputs; their combinations only possible through the successful function of a genetically and environmentally derived and programmed brain. Most of the new combinations prove unimportant and are not selected by society for transmission. But highly productive new ideas are selected – "newness" possible through new interactions that can only occur given the past inputs. No lesser luminary than Einstein subscribed to this view: "The psychical entities which seem to serve as elements in thought are certain signs and more or less clear images which can be 'voluntarily' reproduced and combined... this combinatory play seems to be the essential feature of productive thought..." The new idea $E = mc^2$ was only possible because Einstein's environment had programmed him with information about energy, mass, the velocity of light, and mathematical manipulations. Combinatorial play involving massive numbers of neurons then led to "newness in the world". Originality is not a mystery.

Some have tried to save the notion of free will by invoking events for which they believe the inputs do not uniquely determine the outcome. Chaos theory is invoked. But, as indicated earlier, here cause does determine effect; it is simply that the input is often not known with sufficient precision to predict the outcome. Quantum theory is invoked to allow a range of outcomes from a specified input. But uncertainties in inputs necessarily make for more uncertain outcomes. Uncertainties[2] of output from a specified input into the brain – now much studied[3] – rather weakens the case for free will.[1,4,5]

Religious leaders, journalists, news presenters and politicians give their views through the pulpit, newspapers, radio, television and the machinery of government. The different environments through which these individuals have passed lead to the expression of remarkably different views. For example, a vicar made the following comment at the funeral/memorial service of the BBC presenter Nick Clarke:

> *Terrible though it is to us, God grants the same freedom to cancer cells that he grants even to the most noble and virtuous of us.*

Nick Clarke's former colleague John Humphrys criticised the vicar's comment on the basis of two points.[6] First, that it suffered from "sheer insensitivity of the timing", and that "if a man has died a terrible death from cancer, his mourners do not want to be told that God specifically enabled the cells that killed him". Few would disagree, and many warmly agree, with this sentiment.

Second, John Humphrys commented that (i) "There is also something absurd about the notion that cancer cells can choose what they do", and further added (ii) "What they do not possess in any accepted sense of the word is freedom", and (iii) "Freedom means choice". In these three remarks, he leads us into deeper waters.

First – "There is also something absurd about the notion that cancer cells can choose what they do." At the most fundamental level, in any system in which two or more pathways could be followed and, among these pathways, one is preferred, a choice has been made (the OED definitions of "choice" include "preference"). Cancer cells have evolved to have certain properties, and the choice made – or pathway followed according to natural law – lacks any flexibility. They have no freedom in the paths that are preferentially followed, and the expression "*can* choose" is indeed absurd for alternative pathways have been precluded by natural law and selection. Thus, the second statement "What they do not possess in any accepted sense of the word is freedom" is, like the first, spot on the mark.

Which takes us to the third statement: "Freedom means choice". This statement can now be seen to be false, for choices are constrained by evolution and environment. As such, the possible choices are the antithesis of freedom. It was the vicar's exposure to too much uncritical thought about freedom that led him to the utterance "Terrible though it is to us, God grants the same freedom to cancer cells that he grants even to the most noble and virtuous of us". Ironically, in this statement, he is a victim of his environment and, as such, is unfortunate if censured for his remarks. But a more informed analysis by some, or an emotional analysis by others, ensure that he will be censured.

Social complexity makes for choices of enormous diversity. But the constraining rules do not permit freedom. All biological systems, from cancer cells or complex organisms, are constrained to avoid some behaviours, and promoted towards some others – more likely to be a suicide bomber if growing up in some pockets of the Middle East; perhaps to end up in the US military in Iraq or Afghanistan simply because you met a seductive recruiter outside a supermarket while unemployed; more likely to end up in conflict with Tutsis because you happened to be born a Hutu (and vice versa). A plethora of choices is very different from unconstrained choices. Free will is a manifestly false notion.

There can be no resolution of the contradiction that we are moulded by forces outside our control, yet the success of our social group requires that we act as though responsible for our behaviour. Hence, the media have an endless store of material for their "moral maze" discussions. The efficient function of social groups requires a widespread belief in the false doctrine of free will. As Isaac Bashevis Singer wonderfully said: "We have to believe in free will. We've got no choice."

Programming

Society attacks early, when the individual is helpless.

– B F Skinner

It may be said "a man jams his hands into his pockets to keep himself from biting his nails". As pointed out by the famous behaviourist B F Skinner – one who famously denied free will – this statement implies that one self (a man) is controlling another (himself).[7] This is a rather bizarre notion. So, why is it used?

A word evolves, and becomes a part of vocabulary, because it serves a useful function; the more useful its function, the more frequent is likely to be its use. One of the most useful and interesting words is "I". Children begin to use "I" by about 1½ years, and it becomes the word most commonly used by young children;[8] very important in social relationships and skills, and imbibed from the early environment; an outcome of the functions of the conscious brain. Its implications are diverse[8] and often appear to represent choices that carry little restraint – as in "I am going to bed". But there are big restraints – the necessary physiology, dictated by evolved molecular functions, common to all healthy individuals.

The actions of "I" are triggered by functions in the automatically constructed and programmed systems of the brain, as evident from MRI scans. These establish that defects in the brain cause a reduced, or even very diminished, "I". Therefore the independence of mind usually inferred in the use of "I" implies a falsely based freedom. Sometimes "I forced my brain to do it"; more sensibly, the converse "my brain forced I to do it". Also the much more common usage "I changed my mind". This is also a bizarre statement, because the molecules of brain, which allow the functions of mind, control the actions of "I". The parts of the brain that confer a multiplicity of conscious functions figure prominently in the control of "I". Awareness usually remains even as other functions weaken. Consequently, activities in some parts of the brain allow awareness of failing functions in other parts. At the other extreme, persisting vegetative states result when brain damage removes detectable awareness, but the brain stem – more ancient in its origins – still plays its role in breathing, heartbeat and metabolism.

So why did the concept of the responsible "I" evolve? Because in conscious individuals who live with language in social groups, the falsehoods of free will and responsibility are required for control. Through these falsehoods, survival is served. The illusion of free choice must be both strong and conserved. Language necessarily evolved "I did it" and "you did it" – rather than "it was done"; evolved so that it was difficult to escape from the concept of responsibility. The question "If the brain does not work quite so well, who is to blame – you?" focuses the argument. The answer "yes" is clearly nonsensical, but it is the view held by those who believe in free will. Blame, necessary in the interests of survival, is handed out – "you acted selfishly", "she behaved despicably", "he is the worst possible type". Typical attitudes are nicely encapsulated by the statement made by a friend of the actress Vivien Leigh who sometimes showed antisocial behaviour in the later part of her life:

> *Alas, the disease (manic depression) was so ill-understood in those days that people just assumed she was behaving badly.*

Extrapolating from this thought, those behaving badly can also be regarded as suffering from a disease. "Disease" blurs in a continuum to "condition"; society sits on the horns of a dilemma.

To quote Skinner: "If you're old, don't try to change yourself, change your environment." This quote is an excellent illustration of the way that what comes into your environment – here, Skinner's quotation – sets the menu; the choices that can be considered. And the sentiment is not just relevant to the old – witness the mantra ascribed to the Jesuits – "Give me the boy and I'll give you the man." Although an exaggeration since genes are ignored, the mantra illustrates the powerful environmental influence on what we become.

New knowledge of brain function, in conjunction with the relatively benign environments of some modern societies, has led to the idea of "responsibility" to be put aside in some cases. In a case in the UK courts, the correlation of a woman's violent behaviour with circulating progesterone levels was accepted to mitigate the offences. The violence showed a monthly cycle, corresponding to the automated changes in the progesterone levels associated with ovulation. Therapeutic lithium and a whole host of organic drugs are known to remove, or reduce, antisocial behaviour. Brain surgery can also on occasions confer benefits. These examples, implying the sequence "molecules –> behaviour", are selected from hundreds of similar cases. The conclusions come as no surprise and support the case of brain as chemical machine. With increasing knowledge, the obvious is not ignored in enlightened societies: "human rights" burgeon; "responsibility" is pushed aside in increasingly complex cases.

The negative side of the coin is that opportunistic and selfish behaviour can then burgeon. So individuals must still most commonly be judged "responsible"; without this necessary presumption, mayhem would ensue. With the exception of those suffering serious mental impairment within educated societies, we must be programmed to behave as though possessing free will. But, with regard to mental functions, the variations from "obvious" to "subtle" impairments are stretched along a very long string. For those along many parts of the string, judgemental intolerance is a necessity that is also a tragedy. Fools not suffered gladly, despite the fact that their state blurs in a continuum to mental illness – the latter ingloriously ill-defined.

A call to "rise above our circumstances" is based upon a falsehood, for once memorised, the call is part of our circumstances. Nevertheless, if the mantra of "positive thinking" becomes part of an individual's circumstances, it may – if the proffered dream is realistically attainable – act as a powerful force for change in a desirable direction; stressing the positive can cause positive physical change.[9] But in complex societies, most of us are unable to carry out a reliable cost/benefit analysis. So the proffered dream may be based upon falsehoods: "positive thinking" then becomes dangerous propaganda – "Buy our product – you're worth it"; "APR *only* 16%"; "Bet and *win*"; "The greatest *ever*". Once associated with a glamorous image in the public mind, a trashy product can, through the trait of social conformity, become a global phenomenon. Assertive but nonsensical jingles grab markets because non-testable assertions are widely accepted.

Governments, institutions, advertising and the media use "positive thinking" to their benefit. When the military of any nation seeks recruits, "defence of the realm" may figure; the excitement of the young in controlling powerful, hi-tech weapons is displayed. No recruiting advertisement could display the possibility of losing one or more limbs because it would be counter-productive. An unbalanced stress of the positive is a part of our condition. Global education in rational thought would be required to reduce its pernicious influence; so far, a pipe dream.

It seems bizarre that a creature in which the automated behaviours of molecules rule the beating of its heart, the healing of its wounds, its brain, nerves, senses – in sum, its every function – can be induced to regard itself as free. But preservation of the myth of freedom is our most necessary delusion; so that, should we step outside the rules that promote the success of social groups, the charge of irresponsibility or immoral behaviour can be levelled against us. The evolution of social groups demanded the evolution of new constraints; called for this new programming; required the evolution of the notions of responsibility and morality.

Morals

...The Origin of Species introduced a mode of thinking that in the end was bound to transform the logic of knowledge, and hence the treatment of morals, politics and religion.

— John Dewey in 'The Influence of Darwin on Philosophy'

In the Old Testament of the Christian Bible, there are ten "commandments". These ten are only a portion of the commandments of the Qur'an. There is overlap of the Ten Commandments with those of Judaism — their original source. But in Judaism, the requirements are more rigorous — all 613 commandments given in the Torah should be followed. Concise statements of the ten Christian commandments are:

1. *I am the Lord thy God, and thou shall have no other gods before me.*
2. *Thou shall not make unto thee any graven image.*
3. *Thou shall not take the name of the Lord thy God in vain.*
4. *Remember the Sabbath day.*
5. *Honour thy father and thy mother.*
6. *Thou shall not kill.*
7. *Thou shall not commit adultery.*
8. *Thou shall not steal.*
9. *Thou shall not bear false witness against thy neighbour.*
10. *Thou shall not covet anything that is thy neighbour's.*

Why do such rules exist? As humans engaged in more complex and larger social groups, there occurred a co-evolution of constraints for cooperative behaviour. The resulting constraints — seen here as the written commandments — illustrate moral values as the rules that evolved to promote the survival of these particular aggregates of matter. In this context, commandments 5-10 emerged to promote cooperative behaviour within the group consisting of a family and its neighbours. Rules 1-4 evolved to support a specific religious belief. Such beliefs not only provided hypotheses to rationalise existence. They could — and continue to — give hope in a world that is tough; but are not required to come down from the mountain on tablets of stone.

Religious images and beliefs were inevitably mired in pre-mediaeval superstitions. But they were selected because they enhanced group cohesion; promoted behaviour sometimes humane. In times when faith was essentially universal, the concept of a God with wondrous powers also had the advantage of encouraging a few to attempt to understand "his" physical world. Authoritarian rule from the top was enabled, and provided a restraint on the opportunism of the vast majority of individuals who were lower in the social order.

But the record also reflects the barbarisms of the primitive societies within which the rules evolved. Therefore, injunctions from "on high" are on occasions also barbaric. In Exodus, immediately following the text containing the commandments, God instructs Moses of the judgement to be set upon a servant not willing to comply with God's instructions with regard to potential release from slavery: "His master shall bore his ear through with an awl." It says much about the human condition that a book containing such recommendations from an alleged external intelligence is not only still pressed upon many of the world's inhabitants but also accepted by many of them.

The last five of the Ten Commandments reflect *Do unto others as you would have them do to you* – selected as "The Number 1 Commandment" in a 2005 poll for one of the ITV channels in the UK. This maxim can serve social groups, institutions and individuals by discouraging mayhem. But when individuals act to represent the interests of societies and institutions, they must often deny it. For, if followed, *Do unto others as you would have them do to you* would banish war; would restrain a boss from firing an employee for purely economic reasons; would avoid a used car salesman giving misleading information; would prevent the "cultures" *via* which corporations may maximise their profits through deceits. So, while individuals express at one instant hope for the ethic of the "Number 1 Commandment", institutions often have different needs and programme the same individuals to behave accordingly. Being reminded of the evolutionist's view of moral values as the rules that evolved to promote the survival of particular aggregates of matter then it becomes understandable why moral rules become bent according to circumstances; the moral of "Thou shall not kill" inconsistent with a government's cry for "a moral war".

A judge of several generations ago noted that people were not hung for stealing horses, but to stop horses being stolen. The observation is nonsensical because people were hung for stealing horses; but enforcing cooperative societal behaviour was indeed the evolutionary origin of the action. A seventeenth-century starving boy, having been taught by his father to steal horses as a necessary survival strategy, is sentenced to death by a judge and executed. By the criterion that it is moral to promote the cooperative behaviour that serves order in a society, the judge's actions were moral. But by the criterion of morality *Do unto others as you would have them do to you*, the adolescent acted immorally; by the same criterion, the judge acted immorally.

The morals which constrain humans to cooperate at various levels are but one aspect of a wider rule that flows from the physical, through the biological, to the behavioural sciences: *larger entities are stabilised, and an increase in their population thereby promoted, when there is cooperation between the smaller entities that give rise to them* (chapter 11). Thus, such moral rules are inevitable within social groups of humans. They are transmitted through culture. But they are also programmed in the genes for, if some weak moral constraint has been transgressed, a blush may appear; a strong constraint ignored, sweat runs and hands tremble. The perceived transgression of strong rules triggers a series of automated chemical responses. These responses derive from fear, and represent a later exploitation of this ancient emotion, for fear had survival value prior to complex sociality. Being that accusations of social non-conformity can induce fear, a "lie detector" – a polygraph test – is a very blunt instrument. It measures not truths, but tensions; a test upon which no civilised society should rely for its judgements.

In a primitive and extreme manifestation of fear, spontaneous defecation – well documented in war – is triggered by automated molecular control. Through such effects, dilemmas are compounded, for the cooperative behaviour induced by fear is a tool of tyrants. It is related[10] that Dmitri Shostakovich knew a long-time secretary to Joseph Stalin who, thinking that he had displeased the dictator, soiled his trousers and collapsed; and that Khrennikov, the head of the Composers Union, in delivering to Stalin a selection of names for the Stalin Prize which he perceived were not well received, suffered the same indignity. Dictators can only be told that which they welcome. But how can we know that the author, Solomon Volkov,[10] who presented second-hand these stories perhaps told by Dmitri, was passing on the truth? We cannot know. This is a major problem with many records, and one that raises its head later in this, and in the following, chapter. But the integrated record leaves no doubt that Joseph Stalin was a sophisticated psychopath who used fear as a powerful weapon.

Morality – cooperative behaviour – is a survival tool. It challenges the view of many that morality is a Darwinian trait, increasingly seen as societies became more complex and tightly coupled. As such, it is in competition with selfishness, and therefore sometimes more honoured in the breach than in the observance. So altruism, the trait for which we hope, is required to be a currency of limited circulation.

Altruism

When the missionaries came to Africa, they had the Bible and we had the land. They said, 'Let us pray.' We closed our eyes. When we opened them, we had the Bible and they had the land.

– Archbishop Desmond Tutu

Altruism is behaviour that helps others to survive and/or reproduce at a cost to the helper's own survival and/or reproduction. But all our parental ancestors reproduced successfully in a line that goes back several billion years. Bear in mind that our earliest ancestors (single cells) had much shorter generation times than we do. So, should you not have offspring, you will be unique in a line of a billion ancestral generations – an awesome thought. It is for these reasons that altruism is – as noted by Archbishop Desmond Tutu – a restricted currency. But because cooperation can also promote survival within social groups, it exists and operates at several levels:

1. **Kin altruism** – *sacrifices made on behalf of offspring or other close relatives.* A mother may sacrifice herself on behalf of her child, or a man may die protecting his brother. The basis of sacrifice for a relative was nicely – even if deliberately far too quantitatively – put by the mathematical biologist J B S Haldane: sacrifice of one's life might be expected for one's twin, two of one's children, four of one's grandchildren, and eight of one's cousins.

He did not of course mean this literally. But he was pointing out that the probability of our sacrificing ourselves for our kin is, among many other things, dependent upon our genetic relatedness to these kin. That genetic relatedness can be expressed in terms of the fraction of genes shared with the person for whom the sacrifice is being made: an individual shares half its genes with its son/daughter, one quarter with its grandchild, and an eighth with its cousin.

2. **Group altruism** – *sacrifices made on behalf of members of the same social group.* Classical cases are found in the colonies of ants, bees and wasps (order *Hymenoptera*).[11] The sterile workers of honey bees cannot pass on their genes, but work to the benefit of the colony. Gene transmission – taking place through the queen and the drone that fertilised her – is thereby served.

In human societies, the issue is enormously more complex. Variations in genes and environment cause extreme variations in behaviour, and so group altruisms are expressed to very different degrees.

The potential cost to individuals who act altruistically may be extremely small – as when exercised through the donation of blood. Or it may be high – as when an individual may die while undertaking a dangerous act to save a non-relative. The biological rationalisation for such behaviour is harsh – a hope that the individual making the sacrifice will benefit in the chances of its gene transmission from (i) reciprocal altruism and/or (ii) enhancement of their prestige, or power, in the social group. Although this rationalisation is cynical, altruistic blood donors do encourage reciprocity; brave soldiers do receive accolades and, on occasions, promotion.

Atypical behaviours are found in clerical celibates of either sex. Mother Teresa – Nobel Peace Prize, 1979 – exhibited atypical behaviour in expending a large fraction of her effort in helping the poor and the weak, and not in trying to ensure a security-giving level of personal wealth. As for all nuns obeying their orders, her genetic instructions died with her. However much we might wish, in times of benign reflection, that the world were full – or, at least, half full – of such individuals, it is not currently so. Genes and environments that produce such heart-warming behaviour are rare.

A UK nurse, Barbara Ryder, donated one of her kidneys to Andy Loudon, a stranger who had been receiving dialysis for two years.[12] She was one of only four people in the UK approved to make a "non-directed altruistic kidney donation". The low number making such donations – even after allowing for the problem of a tissue match – shows such altruistic behaviour is indeed atypical. The donations have significant risk both in the short and longer term, and the rules of the game, however much we may, in moments of benign reflection, hate them, have not allowed them to become the norm.

The trait of kin altruism is stronger than that of group altruism, because the former is a more efficient way of promoting survival of one's own genes. Hence, the guiding principle is usually "charity begins at home". The bonuses received by the world's "top" bankers in 2006-2008 and the exposure in 2009 of the selfish expense claims of some UK Members of Parliament demonstrate "charity begins at home" as a strong and persisting trait.

The biologically derived views of altruism seem to fit the facts of human behaviour rather well: evidenced by the profits derived from the transportation of bottled water, in the 1980s, from Europe to California, while people in North Africa starved despite the presence of literally mountains of stored food in Europe; evidenced by in excess of 100 million sufferers from severe worm infections – largely in the third world – when effective drug treatment is available. Economics rule, for farm animals in Europe and the USA are more effectively protected from worm-based diseases than are humans in sub-Saharan Africa. The drugs necessary for treatment have been made available free of charge by the pharmaceutical manufacturers. These drugs soon appeared for sale in New York.

Benefits of cooperation towards survival put the common altruisms in a less generous light than originally regarded;[7] benefits of competition ensure eternal battles between the two. Exploited in these battles are embellishments and deceits.

Embellishments and Deceits

Those who can make you believe absurdities can make you commit atrocities.

– Voltaire (1694-1778)

Deceits are strategies aimed at promoting survival. They are widespread and ancient – evident in the plants: the orchid that seduces the insect; the sundew that lures the insect to provide a meal. They are ubiquitous in human societies. Remarkably, seven million children "disappeared" in the United States at midnight on 15 April 1987.[13] It was from this night that the Internal Revenue Service decreed that the existence of dependent children – for whom a tax deduction was possible – must be supported by the provision of a Social Security number.

Teachers in the Chicago Public School system "improved" their students' examination results in order to achieve standards set by bureaucrats.[13] The school's future – and thereby the financial security of the teacher – was threatened if these standards were not achieved. Teachers less qualified than the average teacher cheated more frequently; unsurprisingly, for they were likely to be under greater threat.

Entertainingly, a computer analysis of the grades given to multiple-choice questions not only indicated falsifications that improved grades. It also suggested that the cheating teachers were so ill informed of the correct answers they occasionally failed to improve the grades as much as intended. Those faced with targets are ingenious in their efforts to attain them; the "targets" of bureaucrats illustrated as a rather silly instrument.

Statements that are difficult to check as to whether deceitful or otherwise are often used to serve vested interests – as concisely summarised by Mandy Rice-Davies. In the 1960s, Stephen Ward was brought to trial in the UK for allegedly living off the immoral earnings of Christine Keeler and Mandy. The prosecuting counsel pointed out that Lord Astor, at whose Cliveden home high jinks parties had been held, denied having met Mandy, let alone having an affair with her. "Well, he would, wouldn't he?" was her response. We may not know the truth of the matter; we do know vested interests are involved.

Deceits are used in attempts to serve countries: the virulent propaganda of war; cooperation used as temporary sham. In World War II, Hitler cooperated with Stalin to partition Poland; then invaded his ally of convenience. The British Commonwealth and the USA then allied with the Soviet Union against Nazi Germany. The Western powers were so eager to embrace Joseph Stalin that his extermination of millions of Soviet citizens did not prevent his portrayal in the West as a smiling and benevolent "Uncle Joe".

The delicate and fluctuating balance between cooperation and competition is illustrated insofar as once the common enemy – Germany controlled by the National Socialists – was defeated, the Western Alliance and the Soviet Union became enemies. The history of World War II provides examples of man's deep instincts for ruthless competition within, even when there is a dangerous enemy without. R V Jones[14] documented instances not only of the British, US and USSR allies attempting to dupe each other in the course of the war but also of the British war administration committees misleading each other for perceived local gains.

Embellishments and deceits are powerful methods for institutional control. The Old Testament – the Hebrew Bible – contained not only the rules that promoted the success of social groups in the Middle East over two thousand years ago. It also had to be "the greatest story ever told". The archaeologists Israel Finkelstein and Neil Asher Silberman provide a case that it was.[15]

The Old Testament records how, when Abraham was 100 years old, his wife Sarah – then over ninety – gave birth to Isaac. And how, from Abraham's loins, came all the nations of the region. These peoples included all the Arab peoples of the southern wilderness, who allegedly arose after Abraham took his wife's Egyptian slave as his concubine to beget Ishmael.[15] It records how the kingdoms which were, at the time of the record, hostile to Israel were "born from an incestuous union".

Specifically, that Lot's two daughters got him sufficiently drunk to sire their children Moab and Ammon. The Earth's most widely printed story not only included wisdoms and rules that were helpful for the success of social groups; it also depicted behaviours close to those in the soap operas of today.

Additionally, Finkelstein and Silberman conclude – based on the evidence of texts and archaeology – that the stories of Abraham, and the successive generations of Isaac and Jacob and sons, are not consistent with the historical evidence; likewise for the exodus from Egypt, Joshua's conquest of Canaan, and the empires of David and Solomon. They appear to have been invented hundreds of years later.[15] The authors conclude Jericho did not have any fortified walls to come "tumbling down" to the tune of the trumpets; that the Israelites, rather than existing as an already distinct culture that conquered Canaan, evolved from the Canaanites throughout the Bronze and Iron Ages.

The larger-than-life portrayals in the Bible, and its commandments, were important in providing the social cohesion of the Jews – an aid to survival. Religion has its roots in natural selection. The trend continued in the New Testament with stories of virgin birth and resurrection from the dead. And, if there are today benefits for survival from these beliefs that are greater than their associated costs, they will continue.

Some widespread human beliefs – whether true, false or embellished – have been termed memes[16,17] and likened to "viruses of the mind". Insofar as they replicate as they pass from one brain to another – sometimes astonishingly quickly – and also evolve, an analogy with genes is striking. They may be styles of clothes, songs or hair – commonly short-lived, but available to be regurgitated through memory; or so facilitating of survival during the last two millennia as to be persistently expressed.

But some embellishments are challenged by the new knowledge of science; false memes perhaps ripe to be displaced by new truths. For how many in future will find plausible the provision of a Y-chromosome – containing approximately sixty million base pairs – to Jesus Christ by miraculous means? Perhaps many, for the question "Why does the omnipotent and loving God permit Auschwitz, destructive tsunamis, earthquakes and storms?" is still debated. The answer "There is no omnipotent and loving God" is consistent with the evidence. But the debate goes on, for harsh realities are often denied by stories that give hope. Fairy stories are much loved by children, but adults have their equivalent – either literally believed, or as Hans Christian Andersen implied, professed in the interests of survival.

The Emperor's New Clothes

In the fairy tale of 'The Emperor's New Clothes' by Hans Christian Andersen, a couple of crooks offer to a vain Emperor clothes that are made "from the most beautiful cloth". But they relate that this cloth is invisible to anyone who is unsuited to his position or, worse still, stupid. Minions of the Emperor are sent to "view the cloth", because he fears being unable to see it. The minions see nothing, yet, fearing they may be judged incompetent, praise the cloth. So when the Emperor makes a procession through town, he is "dressed" in the clothes produced by the crooks. "But he has nothing on!" yells a small child. Then even the adults have a chortle, but the Emperor aloofly continues his procession.

The innocent child tells the truth; the adults hide behind the ludicrous falsehood until it seems safe to step outside it. The story pinpoints the basis of the old adage "Only drunks and children tell the truth". The adults cannot avoid taking on board the experiences of a lifetime; are aware of the dangers inherent in challenging the social order and its prevailing wisdoms – however ludicrous they may be. The trait of conformity, through which truth is often denied, was already demonstrated experimentally in the 1950s;[18] its costs and benefits, and other human traits, are illustrated next by history and some institutional practices.

13

Humanity Illustrated

To illustrate human behaviour over the centuries, some vignettes from human history are considered. Features of the law, economics and politics, and some functions of punishment, deference to authority, and the arts are presented. The focus illustrates human behaviour and institutions as consequences of evolutionary pressures. These pressures lead not only to our warmth, comradeship and sense of fun, but also to our less admirable traits.

History

> *I mean not to accuse any one, but to take the shame upon myself, in common, indeed, with the whole parliament of Great Britain, for having suffered this horrid trade to be carried on under their authority. We are all guilty...*
>
> – From an account of William Wilberforce's speech for Abolition of the Slave Trade, delivered in the House of Commons in 1789

Wilberforce was unusual in addressing a horror most others did not wish to countenance. And, where there is horror, history is frequently written with a slant that whitewashes the record. Those who relate events typically "hype" their texts – witness the press and television. In the records of wars, much is written by the victorious – commonly a record that glorifies their deeds and debases those of the losers. Where history is written by the losers, they usually remove these distortions and introduce their own.

Distinguished historians embrace the view of distortions, exaggerations and even inventions. Moses Finley commented, "But ancient historians, like historians ever since, could not tolerate a void, and they filled it in one way or another, ultimately by pure invention."[1] In providing accounts of the great fire in Rome in AD 54, the chroniclers Tacitus, Suetonius and Dio Cassius confused the situation so effectively that no one has been able to unscramble it.

Therefore, it is frequently difficult to know the truth of the matter.

"Unfettered scholarship" is of enormous value. But scholarship is rarely unfettered, since scholars rely on documents that may themselves be distortions of truth.[1] Records of ancient history must be regarded with a more sceptical eye than those of recent history. First, because ancient histories have often been rewritten many times; and second, because the ancient imagination was, in times of enormous ignorance, capable of conjuring up, and believing, almost anything. In sum, the dictum of the famous German historian Leopold von Ranke to the historian to tell it *wie es eigentlicht gewesen* – how it really was – is always difficult; increasingly so as the events become more remote. Given these caveats, what does history indicate to us about human behaviour?

The horrors that result from the rules of biological selection are illustrated by wars. Going back a mere thousand years, Christianity – although born in the Middle East – was dominant in Western Europe. The Middle East had also given rise to the Islam, and this faith exercised power in much of this region. The Christians, resenting this power, undertook four major "crusades" in the period 1095 to 1205 to bring their "holy lands" under western domination.[2,3] The last of these provides a clear example of human greed and duplicity.

The crusaders required massive funding, not least because they had to pay the Venetians for ships and transportation services to the eastern end of the Mediterranean. The Venetian leader, Doge Enrico Dandolo, suggested that the funding crisis could be alleviated if the Venetians and the crusaders attacked the city of Zara, about 165 miles south-east of Venice on the Dalmatian coast. The perceived advantage to the Venetians was not simply a source of wealth and supplies but also the removal of a competing economy. Zara was duly besieged and then ransacked. The Christian crusade commenced with the sack of a Christian city.

Later events became even more bizarre. A Prince Alexius, maintaining that the Emperor's throne in Constantinople should rightly be his, promised funding and additional knights to the crusaders if they would give him support. As a consequence, the crusaders ended up sacking Constantinople – then the most populous city of Christendom. The crusaders never made it to Jerusalem in significant numbers, but were successful in sacking two Christian cities.

The sacking of Constantinople left a deep and long-lasting wound in the relationships between the Greek Orthodox and Catholic Churches, which Pope John Paul II attempted to heal in the summer of 2001. The Pope referred to the crusaders as setting out "to secure free access for Christians to the Holy Land". But the crusaders had departed from Europe armed with the engines of war; "free access" was not to be obtained by cooperative consent. Social behaviour had budded only locally, and fierce competition was commonly exercised over larger distances.

If genes and environment together determine behaviour then the bizarre combination of gentleness and brutality should not simply be a property of the mediaeval mind. For, over a period of only a few centuries, profound genetic change has not occurred, nor have uniformly benign environments evolved. Plenty that has occurred in the last 600 years shores up this view. In the fifteenth century, the renaissance was well underway in Italy. It not only marked a flowering of culture and science; it also involved savage infighting. The schism that occurred in the Catholic Church in the sixteenth century, resulting in the birth of the Protestant religion, led to a plethora of burnings and counter-burnings.

Concurrently, the export of slaves from Africa, initially largely due to the efforts of the Portuguese, was beginning. During the approximately 400 years up to the middle of the nineteenth century, this slave trade is estimated to have involved the transportation to the New World of around ten million Africans. In the eighteenth century, several nations of Western Europe were involved in the transportation of around six million slaves, with Britain being one of the nations most heavily involved. That conditions were bad for all on these voyages is reflected in the citation of death rates higher than 13% among non-slave passengers and crew. Not surprisingly, the death rate among the slaves during transportation was also high – around 13% is commonly cited, but in some cases an appallingly high 50% is suggested.

The wretched lot of the slaves once in the southern states of North America is well documented: whipping – of both sexes – common; female slaves often suffering rape. During the course of the nineteenth century, these practices were radically reduced due to the efforts of notable humanitarians – altruism is in short supply, but fortunately it exists. The expression of more humanitarian behaviour was unsurprisingly correlated with the evolution of more benign environments; and, once nucleated, a constraint of conformity to horror reduced.

Although advancing technology provided more gentle circumstances for some, the role of genetics in dark behaviour ensures that the products of these circumstances will not be uniformly angelic. Those reared in Britain during the might of the British Empire were taught that, for fairness and excellent administration, it had no peers. But the story has many extremely dark blemishes. During the Indian Mutiny in 1857, in response to the massacre of European families by Indians, the British met cruelty with cruelty:[4] women and children burned alive in their villages; a bayonetted prisoner slowly roasted over a fire; Muslims smeared with pork fat before execution.

"Civilised" Europe shed its projected image several times in the twentieth century – most notably in WWI and WWII. In the latter, the horrors are widely promulgated, but the realities difficult to absorb: about six million Jews worked to death, or exterminated, in Nazi labour and death camps; in the Soviet Union, about twenty-seven million deaths – the majority of them of civilians; German cities bombed into oblivion by the Allies.

It is difficult to comprehend fully the most extreme brutalities of the secular Stalinist and Maoist regimes. Non-conformists in both regimes faced a climate of fear: enforcement through hard labour, or even starvation, torture and execution. Perhaps seventy million deaths occurred in the USSR as a consequence of Stalin's brutal rule.[5] According to an almost uniformly dark biography of Mao Tse-Tung,[6] similar numbers of Chinese were slaughtered under his influence; and genocide was practised against the educated under Pol Pot in Cambodia. Since quantification is notoriously difficult in times of cataclysmic conflict, some might consider that the total killed under these regimes is inflated – say, by a factor of two or three. This view still leaves a total corresponding to the annihilation of a major part of the current population of a large European nation.

So, although gentle care of "one's own" is evident, brutality is also recorded in abundance. Looks like a Darwinian view of a large brained species. Selection produced systems in some circumstances warm and loving; in others, inflamed to cruelty. Both traits are outcomes of the struggle to "be around" – a struggle which also resulted in the creation of gods in our midst.

Gods in our Midst

> Andrea: *Unhappy the land that has no heroes.*
> Galileo: *No, unhappy the land that is in need of heroes.*
>   – Bertolt Brecht's 'The Life of Galileo'

Nations glorify powerful figures of their past in efforts to increase social cohesion. The figures may even be those of glorious fantasy, as in Greek and Roman mythology. Where the figures are real and more recent, the epithet "Great" is often appended to their names – Peter the Great,[7] Catherine the Great[8] and Frederick the Great[9] – to name but a few.

What unifies such individuals is that they were ambitious and powerful – successful in war and in controlling those around them – sometimes in very unpleasant manners. In 1710, Peter "the Great" instructed that all the dwarfs in Moscow be rounded up and sent to St Petersburg. There, they were dressed up by the local aristocracy and, often in a drunken state, provided entertainment at a wedding of two dwarfs.[7] Peter ordered the torture of his son, Alexis, and this led to the death of Alexis. His view of entertainment was revolting even by the standard of the times. One "entertainment" involved air being forced into "friends" *via* the rectum with, on at least one occasion, fatal consequences.[7]

In 1762, Catherine "the Great" took an active part in a coup against her husband Emperor Peter III. Catherine became ruler of Russia after Peter was overthrown and soon killed "in an accident". After installation as Empress, Catherine continued the imprisonment of a previously deposed Tsar, Ivan VI, who had occupied the throne from the age of two months to two years. She maintained the instruction, "If anyone should approach Prisoner Number 1 without my express permission, the guards should kill the prisoner and let no man take him alive." In 1764, the guards of Ivan duly acted on this order; he was murdered in his cell at the age of twenty-five.[8]

If "Greatness" should involve some greater weighting of humanitarian characteristics then Frederick "the Great" – Frederick II, King of Prussia – might have some claim to the epithet.[9] He introduced a number of humanitarian practices into Prussia, and welcomed the company of those, such as Voltaire, who espoused civilised values. However, he precipitated conflict with Maria Theresa of Austria – mother of Marie Antoinette – by invading Silesia, and spent most of his life engaged in war. It was he who ordered the first large-scale destruction of Dresden – "the Florence of the Elbe" – in 1760. In sum, in times when behaviour was constrained to a limited degree, the chance of being remembered as a powerful "Great" was strongly coupled to behaviour now deprecated as "not so great".

The "gods in our midst" syndrome is also evident in the creation of saints. And so it was for Thomas More; "saintly" in his opposition to the establishment by Henry VIII of the Church of England – a challenge to Catholic unity. Yet Saint Thomas had a number of people burned at the stake for heresy. Specifically, he branded William Tyndale – translator of the New Testament into English – a heretic and pursued him to the stake in Holland. It may seem odd that the translator was hounded in this way, but the traditionalists felt under threat – the availability of the text in the language of the wider populace removed some of the mysticism associated with the Latin version. The translation therefore threatened the power of Thomas More's Church.

Heads of states commonly adopted – and still attempt to adopt – god-like images. Following their deaths, "god-like" even became formalised as "god"; for shortly after his death in 14 AD, the Roman Senate declared Caesar Augustus to be a god. Monarchs insured their reigns by God-association, as embodied in the Divine Right of Kings. Jacques-Bénigne Bossuet (1627-1704) took the argument to its ultimate conclusion – the view that kings were God's anointed representatives on Earth. It was used to shore up the rule of Louis XIV (1638-1715), "The Sun King" of France. King James I of England wrote, or had written for him: "The state of monarchy is the supremest thing upon earth; for kings are not only God's lieutenants upon earth, and sit upon God's throne, but even by God himself are called gods." In times when almost all were fearful of religious impropriety, the Divine Right of Kings represented the ultimate insurance policy.

Napoleon Bonaparte was raised to the level of a god-like figure in French society.[10] Brought up in Corsica speaking Italian, he had a meteoric rise in the French military, and his energy and skills brought him phenomenal military successes. His administrative successes are reflected in the Code Napoleon (Code Civil), and in the elegance and functionality of parts of Paris. But much of his life was spent in wars that sent large numbers of young French men to their deaths; his march to Moscow showed woeful ignorance of the realities of Russian winters; his military campaigns resulted in a parlous state of France's finances.

The military successes of Napoleon required the British to search for counter-balancing god-like figures. Wellington[11] and Nelson[12] emerged. On the side of truth, Wellington and Nelson, like Napoleon, are suggested by their successes to be military/naval leaders of high competence. But such individuals are then converted by national systems into "gods in our midst"; the cohesion of society thereby enhanced. So, in the 2005 celebrations of the 200th anniversary of the battle of Trafalgar, Nelson's "humanitarian" approach towards his men was emphasised. But Nelson commented that Christmas Day was as good a day as any to hang a man from the yardarm, and flogged half his crew over a period of eighteen months when in command of the *Boreas*. He broke his treaty with French rebels captured in Naples. Expecting that they would be released to Toulon, ninety-nine of them were later hanged. Nelson treated his wife, Fanny, with anything but humanitarian values. In the creation of national heroes, the realities stemming from the harsh environments in which they lived, and the nature of selection, ensure that only part of the story is told.

More recently, Britain found a hero in Bernard Montgomery[13] in World War II. At a distance, it is difficult to know whether he was, or was not, an outstanding military leader. But he was certainly fortunate in the timing of his appointment as commander of the British 8th Army in North Africa. It was August 1942. The battle of Stalingrad – now Volvograd – commenced in the same month. This was the time at which the tide of war turned in many places against the Nazi military machine – the master of the Axis war machine had overstretched it. Additionally, just as the German forces under Erwin Rommel were beginning to face supply problems, the British forces were becoming better equipped. For Montgomery, the time of his appointment could not have been better.

Of American policy in Vietnam, Montgomery commented: *The US has broken the second rule of war. That is, don't go fighting with your land army on the mainland of Asia. Rule one is don't march on Moscow. I developed these two rules myself.*[13] The history of the last 200 years gives strong support to both rules – so self-evident that Montgomery's claim to have invented them provides grist to the mill of those who criticised his ego. He had in abundance the ego and confidence that, justified or not, aid short-term dominance.

The presentation of these records invites the emotional judgements that go hand in hand with propaganda and patriotism – the opprobrium of some French with regard to Napoleon; of some British with regard to Nelson and Montgomery. But here the spirit is not to attach blame; rather to illustrate realities. The flawless chief executive, the super-human sports star, the glamorous figure of stage and screen – short-term success aided in many fields not only by strength but also by confidence and image. The success of alpha males and females in the social groups of some of our mammalian relatives is clearly evident in our own species.

Humans appear to have evolved with both earthly and ethereal gods because the probability of survival of specific social groups was increased – despite the demise of many as the competitions were played out. Those wielding power within institutions eagerly fulfilled the need, and the hagiography, of both self and others, became a fine art. Even whole disciplines became elevated to pronounce judgements of unseemly security – nicely satirised by the tale of the two behavioural psychologists who have sex, following which one says to the other: "It was good for you. How was it for me?"[14] The tendency to promote a god-like image of self is exemplified by the comment of a character in the same novel: a reviewer likely to be Mr Cleverdick, a publicist better characterised as Ms Syncophant.[14]

*We hold these truths to be sacred and undeniable; that all men
are created equal and independent, and from that equal creation
they derive rights inherent and inalienable, among which are the
preservation of life, and liberty, and the pursuit of happiness.*

– Thomas Jeffersonfrom the original draft for the
American Declaration of Independence

We see in Thomas Jefferson's famous dictum the way we would all hope
to be treated if put into a position of adversity. But science and experience
show that humans have evolved to be slightly different; most with
awesome potential to develop marginally variable skills – no one of which
should, in an ideal world, be demeaned by the prejudice of another. And
humans are not independent, but inter-dependent. Their rights are hoped
to be inherent and inalienable, but unfortunately are frequently sacrificed
for some other cause. Most crucially and sadly, humans often exist in very
different circumstances.

Since all these circumstances are determined in the cauldron of
complexity to which individuals are exposed, logically speaking no one
of them deserves less than another. But some circumstances lead to well-
documented anti-social brutes; others produce anti-social behaviours in
otherwise humanitarian individuals. Our sympathies lie with the latter
group, but the law has as its primary (but not exclusive) function the
empowerment of pro-social behaviour, and even these clear victims of
circumstances may suffer in its hands. In a 1989 case in the UK courts,
A had allegedly fired three bullets at **B**, one of which lodged in the
stomach of **B**. Fearing for the safety of his family, **B** refused to give
evidence against **A**. As a result, **A** was acquitted, but **B** sentenced to
eighteen months' imprisonment. In the interests of the social group, the
law prioritised the prosecution of an alleged gun crime, and the "human
rights" of **B** were sacrificed. The needs of society and of individuals are
on occasions quite different. Often there is no universal "common good".

To take issue with the moral high ground is to court vilification. But
ideals are easily professed yet difficult to bring to fruition. For Thomas
Jefferson's famous dictum was not even remotely realised in the United
States during the next few generations. Many countries now proudly
claim that all are equal before the law. However, the high priority of the
courts to preserve the existing social order is seen in the formal customs
and trappings of the courts. In the UK, the judge and the barristers are
bewigged; but since the accused are "innocent until proven guilty", a

satirist might consider the bewigged the accused and the humbly clothed judges. Systems evolved in harsher times are retained in the interests of strong central authority.

Yet it is true that, as the number of interactions increase within societies, constraints for cooperation are strengthened. Therefore, the laws of modern societies – the vehicles through which social constraints are encouraged and empowered – are a complex business.[15] The advance of agriculture, and the beginnings of technology, led to formerly peripheral parts of the social organism growing in value to the core structure. Hence, evolutionary pressures concomitantly wove them into a larger structure. Some devolution of power ensued – in England exemplified by the wresting of rights from King John by his barons in 1215: "We have granted all these things for God, for the better ordering of our kingdom, and to allay the discord that has arisen between us and our barons…"

The further wresting of control from elites – who had levels of opulence in astonishing and vulgar contrast to the lots of peasants and serfs – followed a few centuries later in the French, American and Russian revolutions. The American Revolution led to a relatively stable system, not least because it had few enemies from within. But in France and the USSR, real enemies of revolution existed within, and cooperation had to be in part coerced. Paranoia evolved to the extent that imagined enemies of revolution were created. In France, under the sway of Robespierre, and in the Soviet system, under Stalin, many of these were eliminated. Rapid change produced – briefly in France, more long-lived in the USSR – meta-stable states with attempts to stabilise them through draconian central powers.

The manner in which the perceived short-term needs of a social system, even one allegedly based in "justice and liberty", might lead to a loss of these very values is illustrated by the reaction to terrorism in the United Kingdom in the twenty-first century. A 2004 judgement by the Court of Appeal of England and Wales permitted the admissibility, in domestic proceedings, of evidence obtained under torture from abroad. This permissibility was in accordance with a Bill passed by the democratically elected House of Commons. Thus, those elected democratically, and one group of judges, expressed an opinion supportive of a perceived immediate, selfish national interest. Individual rights once more sacrificed to the needs of the super-organism.

But some humanitarians and logicians were outraged. For history establishes that, under torture, individuals can typically be made to sign anything. Thus, evidence obtained under torture is not reliable. In October 2005, the Law Lords – a cohort of full-time judges – were asked to overturn the judgement. They did so unanimously in December 2005.

So, the judges of the Law Lords supported a more humanitarian policy than did the judges of the Court of Appeal and the democratically elected House of Commons; those in favour of the abolition of the unelected House of Lords perhaps given pause for thought.

The perceived needs of super-organisms cause governments to break the ethos of the very systems they lead. In 2004, the US President, speaking of alleged al-Qaeda leaders and terrorists, said, "...they will find out there is no cave or hole deep enough to hide from American justice."[16] His remarks were not consistent with their detention at Guantanamo Bay, the US base in the south-east of Cuba. The detainees had not been allowed to meet with attorneys; their constitutional rights less than had they been held within the USA. Guantanamo Bay was a "cave or hole deep enough to hide from American justice."

And when administrations pursue their goals, they may flout not only rights that are enshrined in their national system but also in international laws. The Washington Post reported towards the end of 2005 that the CIA operated a clandestine prison system in Eastern Europe and other countries. Prisoners who could not be tortured under US laws had been transported – "extraordinary rendition"[17] – to countries where their human rights were not guaranteed. Although illegal under international law, the US Secretary of State, in a statement in Berlin in December 2005, defended the existence of this system. Military conflict is a very nasty business; winning perceived so important for survival that "the universal rights of man" are quickly suspended. Administrations, governments and nations, like the biological systems from which they are constituted, attempt to win the race "to be around". Systems of law are, because of the nature of the species, adversarial – with the outcome that attempts are made both to find, and to hide, truths.

Economics

> Cecily: *'When I see a spade I call it a spade.'*
> Gwendolen: *'I am glad to say I have never seen a spade. It is obvious that our social spheres have been widely different.'*
> – Oscar Wilde in 'The Importance of Being Earnest'

Economics analyses the production and distribution of goods and services. Although conventionally applied only to *Homo sapiens*, these activities are very ancient. When flowers evolved, more than 100 million years ago, the plants that produced them provided goods – pollen and nectar – to insects in return for their service of aiding the sexual reproduction of the plant.

Wealth derives for a social group when it is more cost-effective than others in providing widely demanded goods and services. Being that human behaviour is based upon that Darwinian principles, the polarisation of wealth remains enormous in the modern world: champagne for a few; lack of clean water and sewage systems for about forty per cent of the world's population. Idealists hope for a redistribution of wealth among all nations. But realisation of the Utopian ideal seems unlikely in the light of the historical record and the competitive nature of life's systems. The latter allows comprehension of the former.

The scientific method increasingly influences the less complex parts of economics. Key causal variables are identified, and successful predictions made. In other cases, the scientific method is applied without success – for a variety of reasons: the number of variables may be too large, or incompletely considered; or some of the variables may fluctuate with time. The last is a major problem for "predictive economics", for the field is influenced by human behaviour, and successful predictions of when changes in behaviour will occur are difficult or impossible. Complexity results in nations making decisions of breathtaking naivety – whether placing the UK pound on the "gold standard" or fixing some European currency exchange rates within a narrow band (the ERM or "European snake"). After some time, these systems inevitably failed for, in the longer term, market forces change the relative demands for all commodities.

The attempt to save the pound within its narrow ERM band was made by a Conservative government. But the problems of politicians in dealing with complex systems are not ones of political hue; rather due to a lack of knowledge. Under a Labour government at the end of the 1990s, the UK sold approximately half of its gold reserves at an average price near to $275 an ounce. By 2010, allowing for inflation, these assets would have roughly tripled their value. The timing of the Treasury-led move could not have been worse – another example of the lack of wisdom in "high places". The observation is not based on that cheap point scorer of hindsight, for history shows gold is at the low end of its price cycle when inflation is low and other assets, such as equities and property, are booming – as they had been from 1982-1999. The selling of gold was based on the assumption that these other assets would continue to perform better than gold. On the contrary, due to the negative feedback that ensues following greed and excess, the value of many of these other assets then collapsed. Gold, not surprisingly, once more became a refuge.

Nothing new in this change, for history had seen it several times before. It was simply that powerful politicians and their advisors ignored, or were unaware of, these lessons. Dissenters are few for, given the weakness of the gold price in the period 1982-1999, they are easily ridiculed – and possibly dumped – as "the people of yesterday"; conformity once more seen as a strategy that can aid survival. The lesson is reinforced by the observation that the politician charged with leading the Treasury at the end of the 1990s repeatedly claimed "no more boom and bust". "Bust" was avoided during his occupancy of this office not because of his wisdom but because of a global boom. Like all great booms before it, its period was finite and a "bust" inevitably followed.

The difficulties of "predictive economics" are clear. There was a sharp rise in the price of oil in the 1970s; a dot.com crash in the year 2000/2001; and major financial crashes in 1929 and 2008/2009. The difficulties of making successful predictions are proven by the descriptions "sharp" and "crash". The majority of economists and financial analysts did not confidently foresee these events. If they had, the "boom to bust" changes would have been less precipitous and pronounced; financial systems would not have moved to such meta-stable states. So precipitous were the changes that the Royal Bank of Scotland moved in short order from "a sound bank" to one reported in 2009 as harbouring £282 billion of "toxic debt" – equivalent to roughly £10,000 for every working Briton – with the UK government holding an 84% stake. These difficulties are reflected in a long-standing joke in academic circles: in economics examinations, the questions are always the same but the answers vary from year to year.

In the course of the financial crash of 2008/2009, politicians inevitably, and usefully, dragged the senior figures of failing and failed banks and businesses to public inquisition. But the inquisitors conveniently ignored their role in the crisis. In the US, it was the politicians – under pressure, of course – who had repeatedly watered down and finally, in 1999, repealed regulations – the Glass-Steagall legislation – designed, following the 1929 crash, to curb excess in financial markets. They had forgotten, or did not know, the lessons of the popular computer game of the 1970s in which populations of rabbits and foxes alternately grow precipitously and then, equally precipitously, collapse.

Financial analysts and institutions bombard those with money to invest with inducements to "invest your money with us". History shows that, in the long-term, money invested in shares and property outperformed money that languished in banks. But in the short-term – nobody knows. Large salaries and bonuses are paid in times of boom. For any fool can make money in a rising market; and the corollary that the same fools will lose money in a falling market. Thus, when banks argue that they must be allowed to pay large bonuses to the "clever people" that made vast profits for them in a rising market, society must recognise that the banks employed similar people who presided over systems that globally lost trillions of dollars in the falling markets at the beginning of the twenty-first century.

The negative feedback from the disappearance of the wealth of "froth" caused recession, sufficiently severe that many governments were forced to increase the money supply. This was a sin that dare not speak its name, and so was announced as "quantitative easing". Printing money is indeed a sin if done nationally on a scale that causes inflation; and is self-defeating if pursued globally. But not a sin if sufficiently constrained to simply alleviate the depth of a recession, for unemployment cannot be allowed to rise without limit. And the constrained printing of money, dare it be said, is a useful way of a limited devaluing of debt; of devaluing a currency so that the nation's citizens boost the home economy at the expense of foreign travel; and find foreign goods more costly to purchase, but home products easier to export.

Statistics indicate that, on average, investment services provide a return similar to those of portfolios that simply contain a wide spread of equities and property. Thus, there is not as much "wisdom out there" as the hype of the analysts would encourage us to believe. Financial analysts and institutions suffer from the predicament – oft denied – that much of their field is too complex to permit reliable predictions. The customers avail themselves of services of limited value because there is a widespread uncritical acceptance of the hype. A lack of understanding gives vulnerability to aggressive and misleading marketing. The same methods are used in a rather different industry – the anti-wrinkle business, which similarly derives a large cash turnover and profitability.

Conformity to the paradigms that currently reward greed ensures boom-to-bust cycles. Plans to bail out the collapsing Greek economy in 2010 did not simply represent, as some politicians implied, rewarding a country whose economic system was based on "idiocy". Idiotic the system may have been, but equally idiotic were the western European banks that lent money to an increasingly fiscally irresponsible Greece. In the absence of a bailout of Greece, western European banks were under greatly increased risk, with associated risks to western European economies. The bailout was largely based upon realism. And, as this debt crisis unfolded, some financial institutions – far from being chastened by their earlier public roasting – attempted to profit by betting on a collapse of the Greek economy.

All this kind of thing had of course been seen before, but the lessons were forgotten. Only a few decades earlier, banks had unwisely lent to weak economies in South America only to find – surprise, surprise – that some later defaulted on their loans. Only a few decades earlier, those wishing to get their hands on "cheap money" had negotiated loans in Swiss francs because the interest rates were lower than elsewhere. Of course they were, for Switzerland is famous for its ability to successfully combat inflation. But the corollary is that, over time, the Swiss franc will appreciate against the currencies of the borrowers who must – tragically for them – repay the loans in Swiss francs. The trap so obvious, but education so limited that history repeats itself, as when Eastern European countries followed the same practice at the end of the twentieth century.

When a strategy that increases agricultural or industrial productivity evolves in one part of the globe, competing societies that fail, or are unable, to adopt this strategy are under threat. The continent of Africa represents, at the beginning of the twenty-first century, the starkest example of this truth. The plan, reached at the 2005 G8 summit, to increase aid for Africa to $50 billion a year by 2010 was judged by one African commentator to be 50 million children too late. But financial aid alone is of little value; evidenced by a fall in average living standards in Africa despite many billions of dollars of aid for, in many African countries, corruption among government officials is endemic. The greater the polarisation of wealth, the greater is the threat of corruption. Once more, the supply of altruism flows through a narrow pipe. And it is narrow in many human institutions. G8 political leaders feast at a banquet while television concurrently produces film of starving children in Africa. Television crews manage to arrive at such scenes with cameras and associated equipment; but food does not arrive in adequate quantity, and millions die.

Current beliefs promote the idea that the political leaders have a "free choice". But the evolutionary evidence summarised here suggests that these leaders are not free. They are, as for all humans, programmed systems with the emphasis laid on survival. If individuals do have free choice then the rich are truly obscene because starvation could be tackled at a manageable expense to them. Historically, and currently, it is a fact that the problem is not solved. The leaders have been programmed by their evolutionary and environmental history to attempt to "win", and not to trust each other. If nation **A** immediately donated ten per cent of its GNP and its technological expertise to tackle the problem then nations **B**, **C** and **D...** might not do so. Nation **A** would be placed at disadvantage; the electorate of nation **A** likely to throw out those altruistic leaders.

Homo sapiens is currently on an insidious treadmill – a burgeoning population; an economic engine driven towards ever-increasing production; and inefficient recycling from a finite pool of materials. The last would disappear if the products of human industry could be recycled with the efficiency of biology. Currently, many of the products of industry form the indigestible garbage – especially plastics – of tomorrow. Although there will be feedback against excess, it is frequently delayed by obfuscation. Which brings us to politics.

Politics

> *Politics in a literary work are a pistol shot in the middle of a concert, a crude affair though one impossible to ignore. We are about to speak of very ugly matters.*
>
> – Stendhal in 'The Charterhouse of Parma'

"Politics is the art of the possible," said Otto von Bismarck, and the UK politician R A Butler took up *The Art of the Possible* as the title of his autobiography. Bismarck's remark is grossly misleading because politics involves so much that is complex, and therefore uncertain for outcomes, that no one can reliably forecast what is, or is not, possible. So here a different definition of politics is used: the art of attempting to achieve a goal of an individual, institution or a society. Since these goals are an inevitable part of the human condition then so is politics. And since politicians are major power brokers in the competition *vs* cooperation of societies, their decisions are crucial.

But since politics, and the definition of wise goals, is so complex, the definition of a "wise politician" is difficult in the extreme. The politics of those who turn out to be fascist or communist dictators is usually recognised as diabolical only with hindsight, a large fraction of a social group having initially embraced their views. Humanitarians take the view that wise politicians aim to produce the greatest good (or greatest happiness) for the greatest number – an aim deriving from the thoughts of Joseph Priestley, Jeremy Bentham and John Stuart Mill. This Utopian view requires that, in making high-stakes decisions, wise politicians consider not only how they, their institutions or society may benefit but also how any actions will affect the other side and their response. It takes on board the J K Galbraith maxim that, on some occasions, statesmen must learn to win by losing; but always work to minimise the world's barbarisms.

The problem for humanitarian politicians is that humanitarian values are often sacrificed in the fight for survival – with the concomitant sacrifice of the humanitarian politicians. Politicians of the Left fight for a better deal for the poor, but then have often lost the battle because some of them were corrupted by power or because a lack of competition resulted in gross inefficiencies. Those of the Right promote the beneficial aspects of competitive capitalism, but often fail to restrain its excesses: successful institutions wiping out their competitors; manipulation of markets; limited rewards for the majority but gross benefits for bosses. There is something to be said for "mixed economies".

In a complex and uncertain world, politicians evolve to survive in part upon bluster – hence the vacuous speeches, readily constructed as a composite from their characteristic phrases:

> *I will give this nation better government – peace abroad, security at home; opportunities for the creative, help for the unfortunate; education for all and fairness in a balanced society. Where there was strife, let us create harmony; where the infirm were slow to receive succour, let us provide the medical attention they deserve. And to our magnificent fighting forces representing our nation on foreign soil, I say to each and every one of them 'thank you from a grateful nation'. For without them, we would be as nothing. My fellow citizens, let us move forward together to create the future to which we all aspire. I know that the people of Utopia want all these things. For this reason, I shall work to meet the hopes of each and every one of you, as we forge a new, stronger and better Utopia.*

Much applause, for this weapon – no matter how frequently regurgitated, and whether wielded from right or left of the political spectrum – is often, in a sea stretching from possible truths to manifest falsehoods, influential. Consequently, democracies are more effective when the voters are educated in the detection of ludicrous hype. But endless optimism remains the world's most powerful, and often disturbing, propaganda – for who dares to speak against the forces of "good" and "positive hope"? The institutions purveying these messages (and most do) can only be judged on their prior record. They sadly often fail the test, but do win the propaganda war.

Power brokers, indeed most members of the species, are not commonly receptive of opposing views. For inheritance ensures competitive natures, and our environments during the first two decades of life programme us strongly. If one side makes a valid point then the technique of the other is often to ignore the point; or to attempt to score with a counter point. Or even not to appreciate the point. The geneticist Steve Jones presented a course of lectures on evolution in Botswana. Afterwards, he asked one student if he now better understood the subject. "Oh yes," replied the student. "I was created, and you evolved." Point made.

The survival syndrome ensures that politics – whether in government, or in other institutions with strong belief systems – is often a confrontational, nasty and illogical business. In the US, supporters of the National Rifle Association (NRA) have used the mantra "Guns don't kill people, people kill people". That the combination of guns and people increases the probability of a lethal outcome is ignored. Those favouring fast cars oppose lower speed restrictions; resort to "It's not speed that kills – it's drivers". That the combination of high speed and poor drivers increases the probability of death is disregarded. An extra child, born following the implantation of a genetically screened embryo in order that a sibling may receive life-saving treatment, is at one extreme lauded as a "saviour sibling"; at the other, deprecatingly referred to as a "designer baby" or "commodity child". There is much evidence of programmed dogma; use of primitive arts to obtain a goal; and an absence of rational discourse.

Since programmed dogmas are strong and persistent, democracy can only be a slowly blooming flower. It co-evolves with a successful economy, stable institutions and a liberal education. History shows these facets require slow evolutionary change; impossible to impose by quick political and military diktat. Passing generations of politicians, and their associated military, should have in mind historical truths. Britain fought the first of its calamitous Afghan wars in 1842. The British agreed to withdraw, and commenced a retreat of 16,000 troops and women and children from Kandahar.[4] The troops were not given an assurance of safe retreat although women and children were allowed through. Among the troops, only Dr Brydon was allowed to reach Jallalabad to tell the tale of the demise of the other 15,999. The Soviet-Afghan war (1979-1989) took the lives of about 14,000 Soviet troops, with almost half a million wounded or sick. And in excess of one million Afghans were killed. The Soviets finally withdrew. At the beginning of the twenty-first century, western nations once more dip their toes in the acid bath of Afghanistan. Modern communications and transport may preclude a single debacle of the size of the first British Afghan war. But in the new conflict, many of the basic lessons of the old are still relevant – the terrain of much of Afghanistan serving *par excellence* for the pursuit of guerrilla warfare.

The politics of survival leads sub-groups to protect themselves by refuting perhaps valid criticisms. So, unsurprisingly, the Ministry of Defence (MOD) in the UK, and the various divisions of the armed forces, play key roles in judging whether the behaviour of the military has, or has not, passed an acceptable bound. And, among nations the UK, is not atypical. Among complex questions that have been recently raised in the UK: was "Gulf War syndrome" an identifiable syndrome for which

212

compensation might be appropriate; were the deaths of young members of the Army – occurring at the end of the twentieth and beginning of the twenty-first centuries in UK-based army training camps – due in large measure to the treatment received from those around them?

Unearthing the truth in such complex matters is difficult; and when vested interests refute responsibility, the difficulties are increased. The Utopian would argue that the sub-group of society that is criticised should not be allowed a major representation in that enquiry. Historically, the opposite has often been the case, typically justified by "our inside knowledge and expertise is required". But, the evolution of societies indicates why the behaviour – whether based upon specious or cogent arguments – occurs. Long-standing sub-groups are around because their existence has aided the survival of the society of which they are part. And, like an organism, they have inevitably evolved strategies for their own survival.

When soldiers tragically lose their lives in foreign wars, they are typically returned to home soil in the role of heroes. But in the frenetic pace of competition, and in the exercise of privilege and image, the earlier needs of these heroes and their families are often neglected. Soldiers' married quarters sometimes suffer from mould and damp – this despite the refurbishment of MOD London offices coming in at a cost of £2.4 billion in 2004 (about three times the calculated contract cost at Quarter 1 2000 prices). Even allowing for a generous effect (25%) of inflation in the 2000-2004 period, relative to the origin contract, the overspend is in excess of £1,000,000,000. The UK's armed forces number about 250,000 individuals. Even in the extreme case that all had families requiring accommodation, and 10% of this accommodation was currently substandard, this £1,000,000,000 overspend would be sufficient to cover a £40,000 refurbishment of all the substandard dwellings.

When the death of a soldier occurred in Iraq in 2003 in temperatures in excess of 50°C, despite his repeatedly telling medical staff that he was seriously unwell, the UK MOD was not cooperative. It argued that soldiers lost their human rights – under the Human Rights Act 1998 – when engaged outside UK territory. In the general case, institutions inevitably attempt to engineer themselves into positions where they can use their influence to promote their survival and functions. But in this last case, individuals, rather than an institution, were the winners; for in 2009, the Court of Appeal rejected the MOD argument.[18]

Grandeur is an integral part of political control for, as with the courtship displays of lyrebirds and bowerbirds, image gives power and has often aided survival. Among the tools of control are medals, national flags, anthems and sports teams – devices that are enjoyed and simultaneously increase patriotic fervour. Grand structures evolved in part to strike awe in the individual, whether through the pyramids of Egypt, the Forbidden City of Beijing (requiring for its construction more than a million workers), Versailles or Blenheim Palace. Beauty is expressed, but simultaneously egos are enhanced and power manifested.

Elaborate ceremony is a further tool. Individuals are typically strongly constrained against interruption of grand ceremony, for challenges to the system of power sometimes led to the demise of their ancestors. Combinations of ceremony, grandeur and deference surround popes, prime ministers, presidents and monarchs. All these devices evolved to generate a cohesion that typically empowers the social group. They often also especially empowered just a part of the social group, or even primarily the individual at the top of the pyramid. Pope Julius II benefited both himself and his office through the painting of the ceiling of the Sistine Chapel by Michelangelo. But this pope was so bent on waging war that Erasmus gave the following portrayal of his encounter with Saint Peter:

> St. Peter: *Did you distinguish yourself in theology?*
> Julius: *Not at all. I had not time for it, being continually engaged in warfare.*
>
> – Quoted in R King, 'Michelangelo and the Pope's Ceiling', Pimlico, 2003

In these facets of social behaviour, institutions and states are acting as super-organisms. They exercise control over their constituent individuals, and thereby have enhanced the survival probability of the whole. But if cooperation is empowered only by devices running counter to individual needs of the manipulated then psychological disadvantages arise – which brings us to the evolution of the arts.

The Arts

> *All art is an imitation of nature.*
> – Seneca, Roman dramatist (5 BC – 65 AD)

In 1945, Winston Churchill asked David Kelly to help in organising his papers and in producing drafts for his epic work *The Second World War*. In a remarkably prescient phrase, he requested David Kelly to bring "cosmos from chaos". We have seen the principles by which cosmos came from chaos, and primates from protoplasm. But it is still a yawning chasm from apes to the arts that are uniquely possessed by *Homo sapiens*. This leads some to the conclusion that the arts are "God-given"; their practitioners endowed with "God-given" skills. So how does evolutionary theory justify the appearance of the arts?

The common property of the arts is that they move the emotions, often delighting the senses. There are clear advantages in being able to temporarily escape the gargantuan efforts that are necessary for survival. The signals for reflection, enjoyment and solace appear to have stabilised human social systems; promoted degrees of individual independence, civilisation and fun. As these signals were selected, the emotion of "beauty" co-evolved. Although something is said later regarding the biological origins of some emotions, science can say little regarding their subtlety; does not remotely understand the complex molecular details that give rise to the perception of beauty.

But social scientists have advanced our knowledge.[19] Quantitative analysis tells that, by and large, what makes a face beautiful – desired – is global; much wired into our genes. But obviously also an input from culture; and because of variations in culture, the arts have a degree of subjectivity; exercise their magic through a combination of innate and environmental influences. So, where the Darwinian synthesis has taken hold, it is concluded that, as in all evolution, the sophisticated structure of the arts has derived from humble beginnings. Complexity built inexorably on what went before. Seneca's "imitations of nature" became more complex and, when moving in impact, were selected for reproduction and transmission.

The beginnings of music, painting and literature make Seneca's point – primitive music imitating the sounds of Nature; primitive painting – say the Lascaux cave paintings in southern France, variously dated as 15,000 to 25,000 years old – as an imitation of its sights; and primitive jottings to record life's experiences. But by the fifteenth to the twentieth centuries, some forms of music, painting and literature had evolved to astonishing levels of sophistication: Cesar Franck's violin sonata; the final trio in Act III of Richard Strauss's *Der Rosenkavalier*; or his *Four Last Songs*. Far evolved from the repetitive "pom-pom" of some Tudor music; even further from the efforts of the individuals who first blew down a pipe or banged a drum.

The arts are wonderful vehicles to illustrate contradictions; to demonstrate, and even overcome, polarisation; to induce warm cooperation while removing rancour. Music in Venezuela illustrates the case: polarisation with oil wealth on the one hand but slums on the other. Yet a great orchestra has been welded together from the underprivileged children of the country. No less are the achievements of Daniel Barenboim in the Middle East – the creation, in collaboration with Edward Said, of the Arab-Israeli youth orchestra; Beethoven and Brahms better than bombast and blitz. Cooperation is generated through an improved environment. But what happens where Barenboims do not evolve and influence; when people fall outside the net of cooperation?

Punishment and War

> *And thine eye shall not pity; but life shall go for life, eye for eye, tooth for tooth, hand for hand, foot for foot.*
>
> – Deuteronomy, 19:21

> *...nation shall not lift up sword against nation, neither shall they learn war any more.*
>
> – Isaiah, 2:4

Both quotations are from books of the Hebrew Bible, or Old Testament. The first promotes brutal competition through severe punishments; the second advocates gentle cooperation. Chapter 11 told why both of these contradictory admonishments are with us: as within social groups of our close mammalian relatives, a fine balance between competition and cooperation serves survival.

In societies where the difference in the quality of life between the groups at the top and bottom is very large, a fall from the top is not to be countenanced. But those at the bottom experience such horrors that they may "risk all". Therefore, fearsome deterrents should evolve – and did: the incarcerations, tortures and executions widely practised in the Middle Ages. A corollary is that less violent punishments should evolve in societies in which inequalities are reduced – as seen in the relatively benign deterrents of Scandinavia in the second half of the twentieth century. And, more widely in West Europe and North America, there is a collective consciousness of the potentially incendiary mix generated as modern mobility brings long-separated cultures onto common ground. Within these societies, there is a consequent trend away from judgemental values – opposition to "*any form* of discrimination" or deprecation of "controversial" subjects; strong promotion of "multi-culturalism" and "diversity".

The strategies have clear potential benefits, but also potential costs. For there must eventually be costs to governments and institutions that do not discriminate against sex-trafficking; lose sane suggestions because they are described as controversial; do not react against cultures that use medieval punishments, or diversity that allows violence. Members of a generation grow up not daring to discriminate or be controversial, or to speak against distasteful cultures and detrimental diversifications. So obvious that it does not need to be said? Apparently not, for when Nicole Mamo tried to post an advert for a domestic cleaner on her local Jobcentre Plus website, she ended the advert by stating any applicants "must be very reliable and hard-working". When she called the Jobcentre Plus in Norfolk, UK, the following day she was told that her advert would not be displayed if including "reliable" for they feared being sued for discriminating against unreliable workers.[20]

The technologies for killing in war advanced at a hare's pace in the twentieth century. Societies reacted through correspondingly rapid changes in culture, but the tortoise of genetic change could not be obviated – the new lived with the primitive. So the balance between competition and cooperation that might best serve survival was, and still is, in flux – Winston Churchill vilified in the 1930s as a warmonger for wishing to face down the European fascist dictators, and Neville Chamberlain initially received as a heroic peacemaker for apparently avoiding a carnage that would reflect the horrors of WWI. Churchill was subsequently re-evaluated as a saviour in the light of the uncompromising opportunism of the dictators.

The instincts that are primitive in the extreme occasionally resurface – the "beast within" sometimes leaps out: as it did in the well-documented horrific acts of the Japanese army in China in 1937, the Rape of Nanking.[21] The horrors include Chinese being used in "live" bayonet practice; nailed to boards and run over by vehicles; hung by their tongues; and disembowelled before executions. Women of all ages were raped in enormous numbers and sometimes then killed with sticks rammed into their vaginas. Estimates of Chinese dead run from 200,000 to 300,000.[22]

The Chinese American Iris Chang[21] suggested in 1997 that, for these acts, the Japanese should be punished through reparations and the Japanese government should apologise. There had been an earlier attempt to extract an apology from Japan to China. But face must be saved, and "regret" saves more face than "apology". The London *Daily Telegraph* reported on 14 April 1989: "Emperor Akihito of Japan told a Chinese leader yesterday of his regret for the wartime suffering caused to China, Tokyo officials said." Hence, reported in the same article, the cynicism: "This prompted speculation that the new Emperor might have exceeded his brief and sounded genuinely apologetic when he received Li Peng, the Chinese Prime Minister, in audience."

At the beginning of the twenty-first century, requests for apologies for past barbarities of nation **A** against nation **B** have become a global culture. On the positive side, reparations may deter future violent acts; apologies give emotional catharsis. On the negative side, old resentments may once more surface, or reparations cause economic instability – a possible contribution to the causes of World War II being the harsh reparations against Germany following World War I.

At a more local level, tribes war against neighbouring tribes. It is a sobering thought that, during the slave trade of Africans to the Americas, it was not only European raids that procured slaves. African warlords also provided their fellow Africans, and on occasions participated in infanticide and cannibalism against their captives.[23] Nations inflict barbarities on their fellow citizens, as Jung Chang and Jon Halliday documented for China in their biography of Mao Tse-Tung.[6] In its past, every nation has something to answer for, so "apology cultures" could employ bureaucrats of any nation for decades.

Cruelty is not a property of some other race; rather a state into which all countries, and some of their citizens, have descended at some point in their histories.[24] Brutality begets further brutality through the desire for revenge – a primitive trait, the horrors of which are encapsulated in the saying attributed to Confucius (551 BC – 479 BC): "Before you embark on a journey of revenge, dig two graves." The old adage "one man's terrorist is another man's freedom fighter" demonstrates the complexities of punitive conflicts. Behavioural analysis indicates that terrorism, in which testosterone-laden adult males under forty are prominent, is due to numerous factors – weighted very differently according to the case. The factors include feelings of powerlessness, frustration and anger derived from injustices, with poverty and inequality before the law as common themes; ignorance; and a tolerance – or even a primitive love – of violence.

Once a group programmed by the above factors controls sentiment, non-violent solutions may be impossible. But if the causal facets of the conflicts are addressed before the group becomes the arbiter of power, the possibility of peaceful solutions is increased. The delivery of education, economic benefits and the removal of injustices – rather than deliver retaliatory bombast, bullets and bombs – can be beneficial. Slowly, programmes that recognise these complexities develop: even amidst atrocities, the UK Government kept open doors to the Provisional IRA in 1996.[25]

First and foremost, this chapter seems to tell of sophisticated strategies for survival; the balance of competition and cooperation[26,27] respectively reflected in an *eye for eye* and *neither shall they learn war any more*. To counterbalance the no-holds-barred competitions of the past countless generations, and to build the warmth, laughter and love of which humanity is capable, we must hope that "good things" come into our environment. Moving novels tell of what can happen when they do not.[28-30]

If still imbued with the idea of an unconstrained choice to be "good" or "evil", read on. It is time to look at what recent controlled experiments, rather than past experiences, tell about our bipolar disorder – from bonhomie, jokes and joy, to the dark side.

14

Humanity Tested

For us to speak with the young becomes ever more difficult. We see it as a duty and, at the same time, as a risk: the risk of appearing anachronistic, of not being listened to. We must be listened to: above and beyond our personal experiences, we have collectively witnessed a fundamental unexpected event, fundamental precisely because unexpected, not foreseen by anyone. It took place in the teeth of all forecasts; it happened in Europe; incredibly, it happened that an entire civilised people, just issued from the fervid cultural flowering of Weimar, followed a buffoon whose figure today inspires laughter, and yet Adolf Hitler was obeyed and his praises were sung right up to the catastrophe. It happened, therefore it can happen again: this is the core of what we have to say...More often and more insistently as that time recedes, we are asked by the young who our 'torturers' were, of what cloth they were made. The term torturers alludes to our ex-guardians, the SS, and is in my opinion inappropriate: it brings to mind twisted individuals, ill-born, sadists, afflicted by an original flaw. Instead, they were made of the same cloth as we, they were average human beings, averagely intelligent, averagely wicked: save the exceptions, they were not monsters, they had our faces, but they had been reared badly...Let it be clear that to a greater or lesser degree all were responsible...

— Primo Levi in 'The Drowned and the Saved'
1988, written shortly before his suicide

In the preceding chapter, a peek was taken into human records, institutions and practices. Both warming and alarming traits are – as in daily newspapers – observed. A commonly elicited response is "a few people are evil". But why are they evil? Why do others search for good? What do controlled experiments tell about human nature?

A famous experiment[1] carried out by Stanley Milgram supports the idea of humans as programmed biological machines. It shows that the vast majority are readily induced to do things normally strongly deprecated. It should give the whole of humanity pause for thought, for it demonstrates how primitive conformist traits appear in response to strong environmental prompts.

Milgram devised a measure of the obedience of humans to authority. He commenced with a situation in which one participant was set up as an authoritarian figure. The other participants were selected from typical male members of the US populace. These "subjects" were led to believe that the aim of the study was to investigate the effect of punishment on memory. Each subject was told that he was to act as "teacher" in a learning test in which the "learner" was to be given an electric shock when the latter made a mistake.

The "teacher" was given a sample shock of 45 volts, and then began his teaching role from a room adjacent to – but isolated from – the one in which he had seen the "learner" strapped into an electric chair. The authoritarian figure then instructed the "teacher" to increase the shock – on a scale from 15 to 450 volts – each time the "learner" made a mistake. In reality, the "learner" was not suffering, but the "teacher" was made to believe that he was by means of recordings: moans were played back at 75 volts, a demand for release from the chair at 150 volts, cries of pain at 180 volts, and ominous silence above 300 volts.

If the "teacher" protested against delivering the next shock, he was told, "You have no other choice. You must go on!" Despite obvious great anguish among the "teacher" subjects, 62% went to the maximum punishment of 450 volts. Even including the minority who refused to obey the authoritarian instructions, the mean maximum level of shock administered was nearly 370 volts. This study was later repeated using high-school students, and no less than 85% of them went all the way to 450 volts.

The minutes of the Wannsee Conference provide further evidence for the conformity of humanity when in an environment that is controlling or threatening. Reinhard Heydrich, second in command to Himmler, convened the conference in Berlin on 20 January 1942. It was attended by fifteen top Nazi bureaucrats, with the aim of coordinating the Final Solution – Endlösung – through which the Nazis would attempt to exterminate the Jewish population of Europe – an estimated eleven million persons. It is unlikely that all the bureaucrats regarded this monstrous proposal as something with which they wished to go along, but there are no recorded objections. Not only, to use the language of

Primo Levi, were the wishes of the top buffoon – frequently gazed upon with adoration – followed, but the chain of underlings then activated was so large that the policy became practice. Milgram's work suggests that among the underlings were the horrified – but they conformed because the instinct for survival commonly dominates all else.

The behaviour found by Zimbardo and colleagues in a simulated prison environment – created at Stanford University, just south of San Francisco – is also sobering.[2] Volunteers – screened for a normal psychological profile – were divided at random to provide fifty per cent of guards and fifty per cent of prisoners. The respective roles of the two groups were reinforced by the provision of uniforms and dark glasses to the first; and of prison uniforms, and identification only by number, to the second. In the experiment, carried out in the early 1970s, the "guards" quickly took on cruelly dominating roles – even though they were not asked to play such cruel roles.

In a popular account of this experiment,[3] it is described how, following early confrontations between the two groups, the "guards" stripped the "prisoners" and sprayed them with fire extinguishers. Within only a few days, the "guards" compelled the "prisoners" to march down a hallway in handcuffs with paper bags over their heads. Such was the speedy onset of the brutal behaviour that the experiment had to be abandoned after six days even though it had been intended that it should run for two weeks.

The Milgram and Zimbardo experiments establish that inhumane behaviour is – given certain environments – readily induced in so-called "civilised" members of the species. Thus, although bad rearing is a common precondition for repelling traits, inhumane behaviour can also be induced when the earlier environment has not been harsh. Given a uniform, "authority freaks" emerge; facing a powerful authority, conformity commonly controls.

It is emotionally repelling to reach the conclusion that torturers are victims of what their total evolutionary history (including very recent manipulation) has been. Its very suggestion invites Voltaire's conclusion: *Qui plume a, guerre a*. Not only because of our emotional repugnance for what they have done, but also because the conclusion can be used to excuse their behaviour, and thereby allow it to increase. So natural selection ensures that fallen torturers will be treated as though responsible, and will have a greater probability of suffering their own violent deaths.

How confident can we be of Levi's qualification: "Let it be clear that to a greater or lesser degree all were responsible"? Apparently not confident, for Zimbardo's brutal guards had a normal psychological profile. Primo's observation that they are "made of the same cloth as we" is borne out by the experiments. The experiments show that, under an authoritarian image or rule, humanity is frighteningly conformist and potentially cruel; the potential of the uniform to corrupt illustrated. The sad videos from Abu Graib and Basra at the beginning of the twenty-first century were predicted by experiment as not improbable.

The manner in which humanity has been forged permits an understanding of the persistence of riots, massacres and brutal wars; demonstrates why the cloth of which humans are made is frighteningly flammable. Their evil deeds are, like their morals, their concepts of "good",[4] and also often wonderfully warm behaviour, consequences of natural selection.

Beings for the Short-Term

Short-term demise precludes long-term survival. Therefore, in the battle for survival, the provision of life-ensuring and life-insuring short-term needs will often relegate other considerations to the horizon. The constraints upon human behaviour of perceived short-term needs is illustrated by an experiment carried out by two psychologists.[5] Some seminarians were asked to deliver a short talk on a biblical theme, one of which was the parable of the Good Samaritan. *En route* to their presentation, each was made to pass a fallen figure in distress. Some of the seminarians were told that they were late in leaving to make their presentation; others told they had time to spare.

Two striking results emerged. First, those seminarians about to present the theme of the Good Samaritan did not help the fallen and distressed individual significantly more frequently than those talking on other themes. Second, of those told they were late for their presentation, only 10% helped the fallen figure. Of those told that they had time to spare, 63% stopped to help. Those without a perceived pressing need commonly behaved kindly, but a perceived short-term need overwhelmed this kindness.

A US space shuttle exploded shortly after launching with the loss of seven lives in 1986. There was a Presidential inquiry team, of which the Nobel laureate physicist Richard Feynman was a member, to search for the cause of the disaster. Richard dipped a piece of the suspect O-ring seal from the rocket into his iced water drink during a filmed session of the hearings. He noticed it became brittle. This on-camera experiment helped to pinpoint exposure of the O-ring seal to low temperatures before launch as the probable cause of the disaster. But, in the enquiry team's final report, the Chairman (former US Secretary of State William Rogers) had a problem in producing a document acceptable to all its members. For, after all, might a report that was too damning on a particular aspect of the program be also damning to the short-term future of the space agency itself? That is a problem of politics.

That the O-ring seal material was defective at low temperature was proven experimentally. That is a problem solved by science. Richard Feynman knew that truth could not be sacrificed for public relations because "for a successful technology, reality must take precedence over public relations".[6,7] It was necessary to counter the "quick fix" of public relations and convey instead a fundamental truth regarding the behaviour of inanimate matter.

Automated Behaviour

The work of Ivan Pavlov (1849-1936) on the conditioning of dogs is a landmark. His experiments showed that, if a bell sounded each time before a dog was given meat, the dog would eventually salivate simply upon hearing the sound.

Automatic prompting of behaviours has now been demonstrated in a wide variety of animals, and operates in humans. In a car perceived to be travelling too fast, the foot of a passenger experienced in driving reaches for imaginary brake pedals; shoelaces or tie in hand, knots are then tied automatically. "Thinking" about such automated functions can cause their temporary loss. Freedom is not observed here; automated functions of brain are indicated. Despite the much-vaunted grace of human consciousness, do we perform unconsciously more tasks than we might at first assume?

Experiments[8] carried out by Benjamin Libet and colleagues are of deep relevance to the human condition. Benjamin showed that, when we are asked to move, about one third of a second *before* we are aware – conscious – of having made a decision to move, events occur in our brains that lead to the movement. In other words, when this action is initiated, we do not consciously decide to do it; the process is automatic. This result is contrary to our deepest prejudices, for it means that what we consciously regard as the present is in fact around 0.5 seconds in the past.

We wish, and believe, that when Don Bradman punished a fast bowler at cricket, Roger Federer made a perfect return of a fast serve in tennis, or Babe Ruth hit a home run, they were exercising conscious reactions to a ball approaching them at around 90 mph. In all these cases, the ball reaches the sports star in around 0.5 seconds. Benjamin Libet showed that a conscious decision to move, as a reaction to the early trajectory of the ball, could not occur. The process of conscious decision-making is too slow. But subconscious – automated – movements take place in around half of this time, and are the ones upon which the reactions to the trajectory of the ball are based.

These stars make their specialised movements much more efficiently than the rest of us because they received some great physiology, and then beautifully honed it through incessant practice – perhaps with a great trainer. But Libet's findings indicate that, when required to make their fastest responses, they move unconsciously. For this reason, many people find Libet's discoveries disturbing. Humans are shown to be less god-like than before, and instead the automation of their physiology becomes the wonder – and why not? For automated physiologies humble and amaze us elsewhere: the fly evades humans in their attempts to swat it; the gibbon amazes with its acrobatics; and the collie humiliatingly evades attempts to corral it in play. Since the notes played in a *presto* piece must be an automated response, a concert pianist may become less god-like. But the beauty in the performance is not debased, for it lies in the perception of the listener; in the perfection gained through repeated practice; and in the creation of a great composer. Wonders come from performance of molecules.

The interplay between automated and flexible behaviour in my brain was evident on reaching Calais from Dover. On transferring from driving on the left in the UK to the right in France, my hand moved – unproductively – to activate an imagined turning signal at the opposite side of the steering column to that in my right-hand drive car. At a completely subconscious level, the brain had correlated the visual image of driving on the right with activation of a turning indicator positioned as in the cars I had driven – several years previously – for three years in the USA. When the automated action was later overridden, flexibility was demonstrated. But not flexibility without constraint for the brain evolved, in both its automated and flexible responses, primarily not to serve freedom but rather to serve survival.

Dr Jekyll and Mr Hyde

In his famous novel, Robert Louis Stevenson beautifully focused the duality of human nature – the apparent anomaly of both moral and immoral behaviours surfacing in an individual according to circumstances. In a modern context, this duality is demonstrated in experiments summarised by Dan Ariely[9] and others.

Rich veins to be explored through experiment are the extreme emotions – fear, rage, lust, etc. These emotions are seen in our close mammalian relatives and appear likely to be ancient; and therefore to harbour primitive components. So, try a comparison of behaviour in a relaxed and in a sexually aroused state.[9] Male students were asked to indicate approval or rejection of sexual practices widely deemed to be "dirty", "immoral" or "kinky". Their approval ratings of the practices were greatly increased when sexually aroused in comparison with their normal state. Mr Hyde had emerged from Dr Jekyll.

There are plenty of states intermediate between the Jekyll and Hyde extremes. The mild side of Mr Hyde learns of the following data. Individuals are prompted with an "anchor number" between 0 and 99, and then asked to price a variety of articles. Those prompted by a larger anchor number priced articles more highly. For example, those with numbers in the range 80-99 recently in mind priced bottles of wine on average just over *three times* as highly as those prompted with a number in the range 0-19.[9] Earlier prompting subconsciously influences later decisions. If the mild version of Mr Hyde becomes a supermarket executive, he now knows how to manipulate customers; conning crucial in sales.

Better and Worse Living through Chemistry

The placebo effect is the phenomenon through which "false treatments" may have a beneficial effect in the treatment of a medical problem.

Benefits commonly come from surgery because the operation has removed, modified or replaced the physically malfunctioning component – but not always. Pain relief was reported by patients who had undergone one particular standard procedure of arthroscopic surgery of the knee for osteoarthritis. This standard procedure was carried out on one group to allow comparison with a second group that had received only placebo surgery (skin incision but no surgical treatment). In the year following the operation, pain relief, assessed by the patients, showed no statistically significant difference amongst the two groups[10] – a classical case of the benefit that may derive when the brain is programmed to expect something good.

What are the physical changes that can cause the placebo effect? The effect is commonly demonstrated through the provision of a saline or sugar solution instead of an active drug. In 2005, Jon-Kar Zubieta and colleagues[11] gave a pain-inducing injection to healthy volunteers who were then told they were to receive a pain-relieving drug. The "pain-relieving drug" was in fact a placebo. The volunteers reported significantly less pain when they received the placebo; benefit from belief in a falsehood. But, most importantly, positron emission tomography (PET) and molecular imaging techniques revealed that the brains of the volunteers receiving placebos released genetically coded peptides that act as natural painkillers – "endogenous opioids", or endorphins. The physical state of the brain had been changed by the incoming signals of positive expectations.

"The findings of this study are counter to the common thought that the placebo effect is purely 'psychological', due to suggestion, and that it does not represent a real physical change," said Jon-Kar Zubieta. This quotation illustrates how "common thought" has got it completely wrong – a fundamental change necessary in the human mindset. Psychology is the study of mind, and mind is a consequence of changes between physical states. Therefore, even "purely psychological effects" are due to physical change, and these changes can improve our lot. Some molecules are so dedicated to our protection that they believe lies.

The brain's response to false positives of image is used to exploit: a beer given a brand name perceived to be upmarket preferred over another, despite the truth that the "upmarket" product was derived from the other by adding only two drops of balsamic vinegar.[9] The brain's response to false positives is used to mislead in the risk assessment of gambling – the addiction that rewards the few, but leads to the demise of many. In this case, it appears that the false positive of "something good may be round the corner", results in options being kept open long after they should be realistically terminated. These behaviours can be judged predictably irrational today,[9] but evolutionists cannot wriggle. In the past, they must have promoted survival. Survival sometimes served through elevation up the social ladder by insisting on "the best"; sometimes, when the chips were down, risk takers being survivors.

The experiments presented in this chapter prove that more behaviour is robotic than we normally acknowledge. The data discredit our comfort zone theories; destroy freedoms. Both history (preceding chapter) and recent experiments (this chapter) inform about our state; demonstrate that there will be difficulties, as well as successes, in efforts to improve the human condition – as now explored further.

15

The Human Condition

All the world's a stage, and all the men and women merely players.
– William Shakespeare in 'As You Like It'

Awed that networks of molecules can tell spiders how to construct webs, how much more humbled then by the emergence, from the same chaotic cauldron, of the molecular machine *Homo sapiens*? A machine that then built another machine in which it travelled to the Moon? A machine an exemplar of which is William Shakespeare whose comedies, histories and tragedies record emotions almost as extant today as when he wrote them.[1] Humbled so much as to justify taking, in this and the following two chapters, extended looks at the factors that have forged the human condition; more on why humans behave as they do; why there is procrastination – even stupidity – in making tough decisions.

"We Will Ensure it Never Happens Again"

In the welter of frenetic activity, the lessons of history[2] are ignored. Unsurprisingly perhaps because accumulated wisdoms in the conscious brains of the current human population of about six billion will be completely wiped out within less than a century. The replacement billions must once more learn the wisdoms from scratch. And, despite wisdoms, survival instincts ensure that cover ups will follow errors. When horrors are discovered, those challenged as responsible not only commonly deny culpability but they also stress the positive; regurgitate the mantra "we will ensure that it never happens again" – only for similar horrors to surface again a short time later.

The sub-prime mortgage crisis at the beginning of the twenty-first century is a non-trivial case of false stress of the positive, of error and of cover up. Many of the world's major banks, allegedly sophisticated financial vehicles, collectively lost several trillion dollars because they sold debt – directly or indirectly – to people, and institutions, in vulnerable financial positions. It was not only that the vulnerable individuals could be charged high rates of interest. It was also that they must be charged higher rates, because a larger fraction of the loans would turn out to be "bad". When the fraction of bad loans, due to negative feedback, turned out to be far greater than anticipated then almost all became vulnerable. Senior bank employees who opposed such strategies were open to the charge of "negative attitude" or "not being a team player" – an offence that could mean the dole queue. As in what follows the election of uneducated dictators, madness reigned. The power brokers of unbridled capitalism had successfully exercised greed – big time. Bank bailouts rewarded the greed; the taxpayer ended up footing much of the bill. Of one thing we can be sure – "it *will* happen again".[3] For legislation may constrain but can never eliminate greed and unwarranted extrapolations from short-term successes.

Through all this, the support of "troubled" banks by the central reserve banks posed a dilemma. For, in the absence of bailouts of the monster banks, the rate of circulation of money would have collapsed; and then so too would have economies and societies. And, if economies collapse, different groups end up at each other's throats. Even when times are good, differences are a loaded topic.

Differences

In 2007, the following spat appeared in a UK newspaper, the *Independent*, under the front-page headline: "DNA pioneer defends the indefensible".[4] The DNA pioneer in question was Jim Watson, co-discoverer of the double helix. The paper stated that he sought to justify "his theory that there is a genetic basis for differences in IQ". The *Independent* reporters stated Watson's view was that: "Black Africans were less intelligent than Westerners." In his *Independent* interview, Jim Watson did not say this – rather: "This is not a discussion about superiority or inferiority. It is about seeking to understand differences."

The generally accepted definition of IQ is a numerical representation of intelligence.[5] But any response to a question aimed at measuring "intelligence" will depend not only on the nature of the question put but also on the environments experienced by the respondent's ancestors, and on the environments in which the respondent has lived. Cultures attempting "the measurement of IQ" inevitably show bias to their own culture; for if a person performs a task frequently, he/she is more likely to perform it well – which is why a professional darts player will calculate the combinations required to finalise a score of 501 more quickly than will an

Admiral of the Fleet; and which is why young chimpanzees, presumably utilising spatial recognition skills ("photographic memory") honed by eons of activity in an arboreal environment, were able to outperform university students in recalling numbers distributed on a television screen.[6] Yet there is a common persistence in judging "X is more intelligent than Y" or "X has a higher IQ than Y".

Because of the inadequate definition of terms, these pronouncements are nonsensical. It was a common nineteenth-century view for Europeans to regard themselves as more "intelligent" than the Aborigines of Australia. But if lost without food, water and transport in the Australian outback, only a fool would wish to be in the company of, in this context, "a stupid European" rather than a wise Aborigine. The European is not well fitted, either genetically or culturally, for that environment; the Aborigine likely to respond much better to the challenges. Physical performance is obviously also influenced by past and present environments. For example, athletic events show that those who have evolved for many generations, and perhaps also still live, in Morocco, Ethiopia and Kenya have recently performed better – on a statistical basis – in middle- and long-distance running events than others. Different past environments caused genetic variation and they made for richness in the tree of life.

Many rushed to judge adversely Jim Watson's wish to understand differences. Ken Livingstone, the then Mayor of London: "Such discredited racist theories seek to establish a genetically based racial hierarchy..."[7] The UK Government Skills Minister: "It is a shame that a man with a record of scientific distinction should see his work overshadowed by his own irrational prejudices." Members of the broader scientific community were also not to be left out of the dispute. Watson's scheduled lecture at the Science Museum in London was cancelled, and the Cold Spring Harbor Laboratory Board of Trustees, where Jim Watson had spent the later part of his career, decided to suspend his administrative responsibilities as Chancellor. Cuttingly, Angela Gunn, described as a technology specialist, was quoted by the *Independent* as saying: "One decade you're accepting the Nobel Prize for discovering something astonishing, a couple of decades later (*sic*, it was four decades later) you're a doddering old crank whose family probably hesitates before letting him play with the TV remote. The passage of time is a cruel thing." It can indeed be a cruel thing; but so can a lack of its passage.

The *Independent* reporters, and the main body of subsequent commentary, conveyed a message that was "politically correct". Jim Watson was in hot water because, at a time when peoples originating in different environments are now living in closer proximity than ever before, it may avoid conflict to emphasise our similarities rather than any differences. In contrast, Jim Watson, in his statement *This is not a discussion about superiority or inferiority. It is about seeking to understand differences,* was speaking about a desire to understand and to reason. Through the understanding of differences, we may in future be better able to minimise sad outcomes. And, as Jim Watson himself noted, genetic differences do cause sad outcomes, as in the case of his own son, Rufus. Suffering from schizophrenia, Rufus is unable to lead a normal day-to-day life, and his father fears that "the origin of his diminished life lay in his genes". The recognition of a probable reality expressed with humanity.

The dispute illustrates our condition – as highly constrained biological machines, targeted towards survival: scoring a point; unable or unwilling to adopt a more rational approach. Difficult to step outside the emotions – love, hate, jealousy, anger, fear, viciousness, gentleness, aggressiveness, humility, etc – an almost limitless list. These tendencies, coded in the genes and expressed to a larger or lesser degree according to the environment of the individual, are the makers of love *vs* war. They account for Bertrand Russell's observation: "It has been said that man is a rational animal. All my life I have been searching for evidence which could support this."

Science and Politics

I saw no point in listening to boring and predictable speeches when I could communicate with first-class minds through books.

– Comment by the philosopher Nicola Abbagnano on Italian politicians after he spent a brief stint on the board of Milan City Council

In 2006, a UK Parliamentary Select Committee on Science and Technology reported on their political colleagues and associated civil servants.[8] A senior academic concluded that the Home Office had distorted findings and manipulated data so that "much more favourable findings overall" were passed into the public domain. The Select Committee pleaded for "a general recognition that changing policy in the light of evidence should be regarded as a strength rather than as a weakness". So why not apply the dictum of John Maynard Keynes: "When the evidence changes, I change my mind. What do you do?"

The answer for many, throughout history, has been to blindly soldier on – and little changes. In the face of disturbing evidence regarding the WMD justification for the Iraq War, the initial reaction of the British Government was not to change its mind. For, almost concurrently with the release of the above report of the Select Committee in 2006, the then

British Foreign Secretary said in the House of Commons, "I have no doubt that there will be a time when we want to learn lessons." And she further warned that agreeing to inquiries now would send the wrong signal "at the wrong time" to Iraq. She urged MPs to remember that: "Our words...will be heard a very long way away. They can be heard by our troops who are already in great danger in Iraq."

These are time-honoured sentiments, for the morale of troops is key in the time of war. And U-turns come very low on the agenda of politicians, and of humanity in general, for losing face is a sin. But there are stupendous dangers involved in not accepting realities; illustrated by Adolf Hitler's refusal to allow General von Paulus to withdraw from the Battle of Stalingrad – arguably the most terrible in human history. The combined Russian/German losses were estimated as in excess of 1.5 million with von Paulus finally surrendering, to Hitler's disgust, with the complete loss of his army. As seen earlier, humans can be deeply imbued with "the power of positive thinking": when logically based, beneficial; when based on wishful thinking, a recipe for some of the world's greatest disasters. When PR consultants and spin doctors spout falsehoods, they may trigger tragedy.

Science is of course not without its politics, and now sufficiently politicised that there is much bad pseudo-science. Additionally, objectivity is a sufficiently rare commodity that it often takes many years to remove false dogmas, and so important new theories and experiments may lie ignored. Unconventional views of systems are viewed as implausible by the "perceived wisdom". When David Edgerton published his book *The Warfare State – Britain, 1920-1970* in 2005, the politically incorrect and unconventional title was so challenging to the programmed brain that it was most frequently referenced as "The Welfare State – Britain, 1920-1970". The prejudice even extends to the software of search engines; for type into Google the correct title of the book, and the software produces the prompt "Did you mean the welfare state?"

Politicians frequently promote scientific work that they think will have a utilitarian end – not surprisingly, for their environment leads them to conclude that this is the best way to improve goods and services. The historical record establishes that this approach is dangerous. For all the following key discoveries were products of blue skies research, and often made by people working alone: Newton's laws of motion, without which there would be no satellites; Einstein's insight for the conversion of mass into energy, without which no atomic power; the discovery of the electron, without which no electron-beam lithography (to produce the fine patterns required in tiny integrated circuits); the discovery of nuclear magnetic resonance, without which no MRI scans; knowledge of DNA, its structure and of a method to determine its sequence, without which no chance to cure genetic diseases.

The list can be made almost as long as one wishes: the discoveries of antibiotics, without which no rapid cures for many bacterial infections; the laser, without which no laser welding of the cornea; monoclonal antibodies, without which an absence of the simple and remarkably sensitive colour tests used in medical diagnosis; radio transmission, without which no radio or television; and Faraday's discoveries in electricity and of magnetism, without which no electrical power in our homes.

At the time of the above discoveries, their economic potential was completely obscure. The masters of research funding may argue that the discoveries would have inevitably followed later; probably so, but at a cost to the speed of human progress. They also often argue that the situation is different today. This will not wash, for key blue skies discoveries continue unabated. The blue skies discovery of IVF was a mere thirty-two years ago and, since then, almost eight million humans have experienced a joy that would otherwise have been denied.

And independent-minded technologists, rather than governmental prescience, fuelled the British industrial revolution. But it is true that the transistor was discovered, in 1947, in a commercial company in a search with a utilitarian aim. The search was for an on/off device – the perfect matching device to the binary code of computing – to replace bulky vacuum tubes. But even here, the solution was discovered accidentally, and the prototype device was fabricated using paper clips and razor blades. The way in which the transistor would change the world could not be foreseen. Even by 1956, when Bardeen, Brattain and Shockley received the Nobel Prize in Physics for their work, the revolution it would facilitate was only just beginning. Computers were transformed through increasing miniaturisation, power and speed.

Gordon Moore, a co-founder of Intel, observed in 1965 that the number of transistors on a chip was increasing exponentially – doubling approximately every two years. And he predicted that it was likely to continue to do so. So prescient has been this observation that it became known as Moore's law. In 1970, there were around 2,000 transistors per chip; by 2008, the industry could achieve 1,000,000,000 – half a million for each one present in the 1970 device. The trend is not anticipated to reach the buffers until around 2020, and devices with a million transistors per square centimetre are now common. In 2008, it is estimated that the average family in a country relying extensively on advanced technology uses about 100 devices containing multiple transistors.

It is clear that fundamental discoveries are not made to order. Informed and enquiring brains make them; industry then exploits them. Politicians rightly push for a technology policy; and should promote public policies that call on the approach of science. But to have a science policy with prescriptive details is to have a policy for disaster. Expressed in another way, the best science policy is no science policy except for enabling the best institutions – a case for elitism if ever there was one. Good government enables good science by facilitating the collection together of individuals who have become gifted specialists in both enquiry and reasoning.

Fundamental discoveries are facilitated by the ability to think for extended periods without disturbance. As Edward Gibbon (1737-1794) remarked, "I was never less alone than when by myself," and, in the *Decline and Fall*, noted, "Solitude is the school of genius." A modern paradigm for funding of science places emphasis on networking, productivity and apparent utility – a formula that, for the reasons outlined above, is disastrous for the originality required for epoch-making discoveries. For how are the individuals who are appointed to grant-giving bodies to "promote wealth creation" to know where the great discoveries of the future are to be found? The historical record tells – they cannot know. And this truth could not have been better proclaimed than by a man heavily involved in politics at the end of his career:

> *There are…things that we know we don't know. But there are also – things we don't know we don't know.*
> – Donald Rumsfeld, US Defense Secretary in the Administration of
> President George W Bush

This statement is often held up to ridicule but, irrespective of one's view of Donald Rumsfeld, it is a truth that humanity should take to its heart. We know we do not now much about consciousness – therefore something important to explore further. But a key to breakthroughs lies in the things "we don't know we don't know". The mavericks can lead us into these meadows, with benefits stemming from a limitation of the influence of business upon science funding.

Science and the Media

PBS in the USA, and parts of the BBC in the UK, are excellent vehicles for communication of the arts and science. But there is another side to the media coin. At the beginning of the twenty-first century, scientists are increasingly encouraged by the political climate to compete through "attention grabbing" and hype. Trivial observations are often promoted by the discoverers themselves, or by burgeoning numbers of science writers, as "likely to lead to" or as "having the potential to lead to" some technological wonder or cure-all. Such ends are achieved only rarely. The world drowns in data, but lacks understanding.

The headline "Chemical in food tins doubles the risk of heart disease and diabetes", appearing in 2008,[9] illustrates the problem. The chemical in question is bisphenol A (BPA), which is widely used as a can liner in the food and drinks industry. A *correlation* was reported between body levels of the chemical and the increased risk of the two diseases. But, in the headline, the chemical was damned as the *cause* of the doubling of the diseases. Yet the same article cited a point by Professor Richard Sharpe that may get closer to the heart of the matter: "If you drink lots of high sugar, canned drinks you will over time increase your risk of cardiovascular diseases and diabetes (I think we already suspect this), and incidentally you will be exposed to more bisphenol A (from the can lining)." Although this possibility is no more than another hypothesis, it is one perhaps more plausibly associated with causality – sugar, not BPA, as the culprit. Yet, this cautionary note, *cited in the same article*, did not prevent the use of the unwarranted headline.

No wonder our ancestors said, and many of our contemporaries still say, "Let us pray", for folly marches on. Misleading media headlines are justified: "must attract attention" or "must be concise"; or may simply reflect a lack of understanding on the part of the journalist. But the damning of BPA with respect to both heart disease and diabetes is not trivial, because both diseases are major threats to human health. Attention is grabbed, but worry perhaps unduly caused.

But there is more convincing evidence that BPA may act as an oestrogenic compound, and concern is heightened since BPA is also found in plastic baby bottles. So, the case against BPA with reference to heart disease and diabetes is much weaker than is the case regarding BPA acting as an analogue of a female hormone. In the latter respect, regulatory agencies in the US, Canada and the EU reached conclusions regarding the "safety" of BPA that were contradictory.[10] Realistic risk assessment is a difficult, emotional and neglected field.

The effects of such unwarranted headlines are not trivial, for they promote irrational reactions against "chemicals". These pass from adult to child, as reflected when the Nobel laureate Arthur Kornberg heard in a supermarket: "Don't buy that, Mummy – it's got chemicals in it." It will be a remarkable day when someone leaves a supermarket carrying anything that does not have chemicals in it, for even the diehard "organic produce only" consumer never eats anything but chemicals.

Some fields are so complex that the suggestion of a specific solution may signify an undue confidence or a political agenda. In these circumstances, the best scientists might aspire to the image implicit in a remark attributed to former President Begin of Israel: "I can never get a straight answer from these scientists. It is always on the one hand this, but on the other hand that." Or they may wish to emulate John Kenneth Galbraith, an economist and former adviser to President John F Kennedy. When in a radio interview Galbraith was asked, following the partial Asian stock market collapse in 1997, "What are the stock markets going to do over the next few weeks?" he replied languidly that he was not in the habit of making predictions with regard to the future short-term behaviour of systems as complex as stock markets.

Complexity ensures that the future short-term behaviour of stock markets is not reliably predictable. But, surely, the average changes in the global climate over the next hundred years should be a prediction within our grasp? Well, perhaps yes; but perhaps not. The strengths and the limitations of the scientific method, and the realities of politics, are beautifully illustrated by "global warming".

Global Warming

He who will not economise will have to agonise.

— Confucius (551-479 BC)

The photosynthesis of the last billion years or so "fixed" atmospheric carbon dioxide and water as more complex molecules in plants – with oxygen as a by-product to provide a new atmosphere. These complex molecules subsequently formed massive amounts of oil, gas and coal. Modern humans have returned the carbon from these molecules to the atmosphere at an increasing rate during the last 100 or so years, with the regeneration of carbon dioxide. So much so that of the total CO_2 now in the atmosphere, about thirty per cent has come from anthropogenic emissions. The level of CO_2 in the atmosphere is now higher than at any point in the last 600,000 years – and the increase has come about very quickly. And CO_2 is effective in trapping the Sun's warming radiation – a "greenhouse effect".

In the discussion of global warming *possibly* due to the above factors, there are four distinct questions that arise successively if the preceding one is answered in the affirmative. First, is warming occurring? Arguments against global warming based on "rubbish, last year we had a cold summer" or "we had more snow last winter than for ten years" are likely to be irrelevant – likely to represent local fluctuations. Data for, or against, warming must be globally integrated; and preferably over at least a few decades – more if possible. The majority view from measurements is that during the last 100 years, the average global temperature has risen by about 0.7°C. But, even here, there are some dissenters, for the determination of average global temperatures is not a simple process. Yet the conclusion of warming convinces many through direct observations: over the last one hundred years, glaciers have receded;[11] the area of the Arctic ice sheet has been reduced; sea levels have risen by about twenty centimetres. This last effect is caused largely by thermal expansion of the oceans, but supplemented by run-off from glaciers and land-based ice sheets. Averaged over the last 100 years, warming appears to be real.

Second question: are emissions from human activities responsible for the warming? Since CO_2 is certainly a greenhouse gas, it might appear assured that CO_2 is a major cause. And the large majority of scientific experts conclude that emissions from human activities are probably responsible for the warming (as summarised in an IPCC report[12]). Critics of the majority viewpoint point to a few clear errors in the report, or to the debatable presentation of one or more issues. But such shortcomings in handling masses of data are typical in human affairs, and do not in themselves negate a large body of evidence.

The critics also point to a more legitimate concern – the "we-are-all-like-sheep" syndrome, leading to the squeezing out of unfashionable opinions. For example, there are data to support thirty-year oscillations between warming and cooling cycles over the last 120 years,[13] such that there was a global cooling of about 0.2°C in the period from 1940 to 1970. If these thirty-year oscillations are considered over the last 100 years then, since this period includes two warming cycles and only one cooling cycle, a part of the global temperature rise of 0.7°C in the century can be accommodated. It is argued[13] that the remainder may also be accommodated without invoking anthropogenic greenhouse gases as a cause, because there has been a rising average temperature of about 0.5°C per century as the world recovers from a cooling cycle that ended circa 1750. In this scenario, alarmist projections are negated.

But the IPCC projections deserve serious consideration. For *if* increased CO_2 levels are a major cause of the recent warming, worries appear. Melting of permafrost and warming of the oceans would add more greenhouse gases (including water vapour) to the atmosphere; and loss of snow cover would result in absorption of more solar radiation.

These three effects and warming are mutually enhancing through positive feedback. Fourth, the deforestation of the globe accelerates, resulting in less absorption of CO_2. The Brazilians are criticised for logging in the Amazon rainforest; the Indonesians for actions on their patch. But they are driven by the same motives as led to the deforestation of much of Europe and North America in earlier times – survival and economic gain. Fifth, the consumption of fossil fuels up to the end of 2007 continued to increase. Due to these variables, and others, there is inevitably much uncertainty in the IPCC predictions – based on computer models – with regard to the possible extent of future global warming. But they are certainly worrying: that within the present century, the warming will have serious consequences for humanity and for the balance of life on the planet. Although nothing is certain at this level of complexity, prudence seems to be the appropriate watchword.

So, third question: what might be done to reduce the worry? Here, there can be no disagreement on the answer: reduce the use of fossil fuels. Such a policy would also have the advantages of extending both the supply lifetime of a remarkably useful commodity and the period over which possible – but reduced – climate change could be monitored. So, the fourth question must be posed: actions by whom, and to what degree? At this point, somewhat uncertain science is superseded by very uncertain economics and politics. In modern economies, short-term success requires energy in copious amounts, and the cheapest and most convenient form of energy is fossil fuels. So far, the selfishness of nations, and fears of the promotion of recession, has precluded action on a scale that would reduce worries.

Politicians paid lip-service to tackling the problem while failing to do anything effective in the seventeen years from 1990 and 2007. Since the so-called Rio Accord (1992) and the Kyoto Protocol (opened for signature in December, 1997, but coming into force only in February, 2005), global consumption of fossil fuels increased by an estimate in the range of thirty to forty-five per cent. In this period, the USA did little that was useful because its administration emphasised that its economic success is dependent on oil consumption; the countries of Europe announced plans to reduce their consumption of fossil fuels, but slightly increased them. The 2009 UN Climate Change Conference in Copenhagen showed the issue being treated by some with increasing seriousness, but with the selfishness of nations still preventing binding agreements.

Green energy generation slowly emerges from the mire, but immediate and marked reductions in greenhouse gas emissions are still a dream. James Lovelock has remarked: "At some stage, climate change will come onto the agenda, probably when it starts killing a lot of people and is making life very uncomfortable." The case for prudence is that, by that time, it may have come onto the agenda too late. Getting CO_2 out of the atmosphere would be a tough call for there is a big adverse entropy problem to be overcome.

The political problem is exemplified by the statement of the UK Chancellor of the Exchequer in September 2005 (at the time of an oil shortage and a burgeoning price): "I call on world oil producers and oil companies now to support the British plan agreed this weekend by the whole international community to raise production." In politics, publicly professed views can quickly change for the same Chancellor was, by October 2006, calling for a long-term framework of a worldwide carbon market that would lead to "a low-carbon global economy". Politicians, and the systems over which they wield power, are primarily concerned with survival in the short-term.

Through all this debate, the desirability of becoming more "energy efficient" is clear. A guide to an energy policy that "adds up" will come into your environment if you read *Sustainable Energy – without the hot air*.[14] As in other fields, sensible policies battle against emotions deeply seated in genetics, diverse cultural views, competition and ignorance. Through the power of the media, ignorance often gains widespread attention while the truths of science are denied. An experienced journalist of the UK *Independent* newspaper gave, in 2007,[15] the following numbers of gigatonnes (billions of tonnes) of CO_2 put into the atmosphere from the specified sources; in astonishing contrast are the numbers (in parentheses) estimated by a well-trained physicist:

(i)   Burning of fossil fuels: 7 (26)
(ii)   The biosphere: 1,900 (440)
(iii)   The oceans: 36,000 (330)

The differences in the numbers (from a factor of about 4 to a factor of in excess of 100!) should strike fear into anybody inclined to believe the "truths" appearing in newsprint. The physicist, David Mackay,[14] identified the errors:

(i)   Confusion between 7 (the number for carbon alone) and the number for CO_2 – the correct number is derived from $7 \times (12 + 16 + 16)/12 = 26$.
(ii)   Flux into the atmosphere from the biosphere is only 440, not 1900 – an overestimate by a factor of just over 4.
(iii)   36,000 gigatonnes is the total amount of carbon in the oceans. The flux into the atmosphere each year of CO_2 is about 100 times less.

The newspaper article was written from the viewpoint of a sceptic: "The oceans send a further 36,000 gigatonnes of CO_2 into the atmosphere, all of which adds up to one big reason why some of us are sceptical about the extraordinary emphasis put on the role of human fuel burning in the greenhouse gas effect."[15] So, numbers for CO_2 "added" to the atmosphere [items (ii) and (iii)] that appear to dwarf human contributions [item (i)] are erroneously selected.

But the errors of the journalist are much worse. For the numbers in (ii) and (iii), even after correction of the gross numerical errors, are not net additions.[13] Although large amounts of CO_2 are added to the atmosphere from the biosphere and oceans, this addition is very largely cancelled out through re-absorption. The evidence for such balanced re-absorption is not hard to find. For, prior to the time the industrial revolution was getting into gear around 1850, atmospheric CO_2 levels had been remarkably constant for 1,000 years; despite the fact that the oceans and the biosphere were very much in operation during this period.

In a complex world, further muddying the waters through false conclusions is the last thing that is needed. Human gullibility leads to the acceptance of false conclusions promulgated by most professions – including those of poor scientists.[16] And the problem is confounded when cultural beliefs based in ancient dogmas oppose new knowledge – in ways considered in the following chapter. Humanity is challenged to think anew.

16

Humanity Challenged

The Message

*Thomas Gradgrind, sir. A man of realities. A man of facts
and calculations. A man who proceeds upon the principle that
two and two are four, and nothing over, and who is not to be
talked into allowing for anything over. Thomas Gradgrind, sir –
peremptorily Thomas – Thomas Gradgrind. With a rule and a
pair of scales, and the multiplication table always in his pocket,
sir, ready to weigh and measure any parcel of human nature, and
tell you exactly what it comes to.*

– Charles Dickens in 'Hard Times', 1854

The railing in the mouth of Mr Gradgrind emphasises degradation
when quantification is used in isolation to describe humanity and its
close relatives. The number of genes is only a part of their rich story, and
arguments against the manifestly absurd view that "we are what are genes
are" have been assembled.[1] Rather, as argued in chapter 12, there is genetic
and environmental determinism.

When genes and environment act to determine our fate, even
miniscule variations in either may have enormous consequences. An
individual on a strange street may turn left, rather than right, even when
there is no perception of different consequences between these alternatives.
But there was some cause of the turn to the left *vs* right, even if too subtle
to discern. Subsequently, a chance meeting, consequent upon the left turn,
may determine the partner chosen for sexual reproduction. So, for as long
as the individual's genes are propagated, they are in combination with
others in a way that was determined by a cause of great subtlety.

Our forms, functions, behaviours and decisions seem to be the
consequence of effects within an incredibly complex, but nevertheless
causal, chain; all consequences of causes of which there is current molecular
memory – some extending back to the primeval soup, others experienced

right up to the present instant. Every memorised event generates a new system within us; occasionally, even single sentences can change lives. In this web of complexity, humans are channelled into different states. Bizarrely different views of the world are held.

Science, Creationism and Intelligent Design

> ...*the eye is today a showpiece of the gradual, cumulative evolution of an almost perfect illusion of design.*
>
> – Richard Dawkins 'London Times Weekend Review' 21 May 2005 (see also his Climbing Mount Improbable, Penguin Books, 1996)

The sceptics of science may emphasise a view expressed by Wittgenstein that "If all the problems of science were solved then none of the problems of life would be touched." Wittgenstein has some cogent facts on his side, for descriptions of the physical and chemical processes that lead to toothache are of little interest to someone experiencing the pain. But he overstates the case. Physical and chemical experiments, and the derived understanding, provided treatments that removed the pain. Some of the problems of life were solved.

Most of those living in advanced technological societies live much longer, and less harsh, lives than did their ancestors. The net benefits of science and technology are unambiguous. Yet, there exists an "anti-scientific" lobby. Part of this lobby points out that scientific discoveries have been exploited in very ugly ways. Dominance on a global scale does often go hand in hand with advanced technology. But it was the efforts of a social group to win a battle that led to the exploitation of $E = mc^2$ to produce atomic and hydrogen bombs, not Einstein. The problem is human nature in its societal relationships.

A sub-group of the anti-scientific lobby, particularly active in the US, refuses to accept the scientific proofs of the age of the Earth, and the origins of humanity. They prefer the biblical view and, in opposition to the true scientific method, have formed their own subject – "Creation Science". A US court ruled that "Creation Science" is not science. The judge defined science as what scientists do; he then found that scientists typically do not carry out and publish research with a view to furthering such a view of creation. Those who do this, with their own publications, are thus simply viewed as creationists. Those imbued with the scientific method will regard the judgement as wise.

Fascinating theatre is provided by the debates between scientific orthodoxy and "creation scientists". In one case, a scientist stated that the difference between the genes of humans and chimpanzees is only a few per cent; surely convincing evidence to support the idea of common origins? The creationist responded that a melon is circa 97% water, and a cloud is essentially pure water. Should we therefore conclude that a cloud and a melon are similar things? This response is much appreciated by an audience consisting largely of creationists. But it impresses only those who are unaware of the enormous information content of genomes. The construction of complex objects – melons, chimpanzees and humans – requires a very large input of information, and that information is remarkably similar for chimpanzees and humans. In contrast, the construction of a cloud requires a much smaller set of constraints.

In the same debate, the creationist raised a laugh from the audience by pointing out that the scientist would have us believe that humans came about spontaneously from hydrogen gas. Well, not quite. But, given a few particles and physical laws, humans and all other organisms have indeed spontaneously evolved – as described in these pages. There is an enormous body of evidence to support this view. One doubts that the creationist has made any serious study of science, and in particular of the Second Law of Thermodynamics.

Suppose you are conversing with a companion who maintains that there is a little green man sitting on his knee. You look but see no green man, and inform your companion accordingly. At this point, your companion gives the proposed green man the property of being visible only to him. The scientific method rejects such argument. First, because it is *a priori* implausible that two humans, known to function in very similar ways, should have very different cognition for the proposed green man. Second, because the proposed property smacks of an arbitrary convenience to thwart the reasoning of the doubter; it involves a property that cannot be tested. In establishing whether a belief is, or is not, credible, science applies the criterion: "Hypotheses not to be entertained – those which cannot be subject to test, especially if contrary to evidence based on careful and reproducible experiments."

But some of the most powerful individuals in the world do not apply these criteria. In 2005, the President of the USA publicly promoted the teaching of "intelligent design" – as might be ordained by a deity – in US schools. In his remarks, the then US President said students "ought to be exposed to different ideas". He was proposing that students ought to be exposed to ideas that are contrary to the testable evidence. The vast majority of a distinguished part of the US scientific establishment opposes such proposals, as judged by the measurement of attitudes among members of the US National Academy of Sciences. The survey found that only seven per cent of these eminent scientists had a personal belief in a deity.[2]

Despite the evidence, some, who allegedly have training in logical deduction, have problems with natural selection. The Archbishop of Vienna wrote in July 2005 in the 'New York Times' that evolution as "an unguided, unplanned process of random variation and natural selection" is not true. And added, "Any system of thought that denies or seeks to explain away the overwhelming evidence for design in biology is ideology, not science." The Archbishop was behaving in a manner that we commonly share: a desire to win the argument. But, having been programmed in a religion that includes miracles, testable truths were not high on his agenda – for science has produced clear evidence for random variation and natural selection; for design, it has produced no evidence.

Religion

I cannot accept the personal God of conventional Christian religion who is working out His purpose through childhood leukaemia and congenital malformation.

– David Weatherall, formerly Regius Professor University of Oxford

God wishes to see people happy, amidst the simple beauty of nature.

– Anne Frank

Anne Frank, who died at age fifteen in the Bergen-Belsen concentration camp just a short time before the end of WWII, expressed in her diaries a view of a loving God. David Weatherall expressed a view based on a wider and longer experience of life. Views vary according to experience, which determine the specifics of belief systems. And, as seen in the preceding section, this experience may result in the expression of views that are in conflict with new knowledge.

Although evidence-based truths have enormously benefited societies, faith is often a basis for pro-social behaviour. The expectation is that pro-social behaviour will increase as the interactions within, and between, social groups increase in number. Consistent with his expectation, the message of the New Testament focuses more on peace and conciliation than does the Old (chapter 12), for with the passage of the centuries, more cooperative constraints were needed. The Christian message – though frequently manipulated for gain – spread like wildfire: endearing cooperation expressed when people gather together in religious buildings; peace and solace provided by ceremony, and by beautiful buildings and music. Tenderness is shown when some adherents care for the mentally ill, or those dying in a hospice; when they care for the weak, diseased and hungry; when they express a sense of love and mutual understanding at rituals associated with births, weddings and deaths. So accommodation between the humanitarian aspects of religion and new knowledge is a desirable goal.

But strongly held religious beliefs often come with horrendous costs. These include violently anti-social behaviour as one dogma sought, and still seeks, to survive at the expense of another. Other costs include the denial of opposing truths; the primitive dogmas of parts of the Old Testament; and adherence to rigid social strata, as found in the Indian subcontinent, where somewhat over ten per cent of the populace has been designated "untouchable". In order to survive, some members of this caste directly carry out the removal of the human waste of others. Although the caste system has been abolished under the Indian constitution, the practice remains. And it extends through other parts of South Asia among various religious groups.

The Darwinian basis for religion has already been indicated (chapter 12). It is because survival probability must have been greater among those with faith – despite the carnages of inter-religious wars – that religions were an almost ubiquitous part of our past. The values of religion that may be useful for survival include social control, the pacification of a brain with regard to origins, and – if the religion includes a message of cooperation – some gentle and considerate behaviour to others. And, last but not least, release from a reality often bleak. For, in times of greater ignorance and hardships (or even today for some, despite the possibility of acquaintance with the facts), the palliative of "Sleeping with Jesus" is infinitely preferable to the reality of the Grim Reaper. Powerless against the overarching need of biology, one need of the species was a superpower to challenge the inevitability of decay; to deny the permanency of death. Therefore, humans have shown a propensity towards religion,[3] and a classical example of convergence among diverse past cultures was found in the need for gods. This propensity is currently attenuated by increased secularism in science-based thinking in northern Europe, but only the future will uncover the long-term global trend.

In male-dominated societies, the assumed external intelligences of religions were unsurprisingly commonly male. In traditional paintings, the Christian God floats on high; through accoutrements that include a flowing beard, is imbued with wisdom. The scholar Jonathan Miller noted that the church paintings representing the biblical story were so ubiquitous in the Middle Ages that disbelief would be almost unthinkable to the medieval mind.[4] The concepts of "heaven" and "hell" powerfully persuade those held in their sway; the promise of paradise *vs* the threat of eternal fire a powerful controller. Where life is harsh, and modern knowledge has penetrated little, paradise is eagerly grasped.

The religion to which a person subscribes is most probably the religion into which the person was born. Conversions exist, but occur with a relatively low probability: Shinto beliefs common in Japan; Hindu beliefs in India; Christians more common in Western Europe and America than in the Near- and Middle- East; and Muslims more common in the latter areas. With high probability, specific beliefs are transmitted within cultures, thus demonstrating the susceptibility of humans to early programming. Faiths are strongly planted memes. Commonly, a specified religion takes itself to be "the only true faith" – if true then all the others must be false.

Web statistics suggest that a high percentage of the world's current population believes in gods of one form or another, or is religious in another sense.[5] Estimated numbers of adherents are: Christianity, 2.1 billion; Islam, 1.5 billion; Hinduism, 900 million; Chinese traditional religion (which includes ancestor worship – belief that deceased ancestors have a continuing existence – rather than an explicit belief in gods), 394 million; and Buddhism, 376 million. The total is circa 5.2 billion, with the total in the Secular/non-religious/agnostic/atheist category put at 1.1 billion. These statistics suggest that more than eighty per cent of the world's population is religious. The percentage may be inflated, for conformity is a strong trait.

Non-believers present three rebuttals against the God hypothesis. First, they challenge the argument "the Universe must have been created by somebody" by asking the question: "Who, or what, created the creator?" If the answer is that the creator does not require a prior creator then there is a logical inconsistency in requiring "somebody" for the creation of one system (the Universe and the Earth's organisms within it) while then casting aside the same requirement for another – a presumed God. If a logically consistent approach is adopted then the creator, an intelligent being, requires pre-existing organising principles – and thereby a preceding intelligence – to account for its existence – and so on. We end up in an infinite regression – the story of the homunculus all over again.

If the God hypothesis is couched in terms of an omnipotent and benevolent God then rebuttal two emerges. For the omnipotent and benevolent, having created a vast number of species, then eliminated the vast majority of them in the five major extinctions. Not much benevolence there. The third rebuttal is that life on Earth never has been, is not and never can be based on universal benevolence. For among animals, survival commonly demands that one organism eats another; the carnivores necessarily bathed in blood at every meal. The existence of war, starvation, disease, carnivores, decay and death, admixed with limited cooperation, rather point to Darwinian selection.

Adherents to God hypotheses may be divided into three loose categories. First, those who believe in the discoveries of science but retain a God hypothesis because the Universe is a remarkable place, and/or they need the hypothesis "to get out of bed in the morning". In terms of the global population, they are relatively few in number. Second, those who are unaware of the strong evidence-based case for the evolution of the Universe, of organisms and of behaviour – a group that comprises a large fraction of the world's population. Third, those that are aware of many of the conclusions from science but do not believe them. Many in this group are unable to come to grips with the idea that something as awesomely complex as a flower, bird or human being can, given the physical laws that govern the behaviour of matter, come about without design.

And how can any human not be awed by the translation, within weeks, of a fertilised egg into that flying phenomenon the Arctic tern, capable of covering 20,000 miles a year? It is not easy to grasp the power of natural selection, nor the timescales involved, nor the basic physics, chemistry and biology. But its power has been promoted over many years.[6,7] The sceptics ask, "But of what use is half a wing?" In fact, something much less sophisticated than "half a wing" confers considerable advantages. Flying fish have enlarged and modified fins. When chased by a sea predator, they take to the air for distances up to twenty metres. A bat wing consists of fingers evolved to a great length, and skin flaps to cover them from base to their tips; a structure that allowed them to become the only mammals to join the birds in the skies. The sceptics should heed the variation produced from wolves. Within the last 100,000 years – and perhaps within 10,000 years – selection produced the very diverse forms of the Chihuahua and the Great Dane.

We are here because of remarkably few immutable laws. The changing of water into wine (if other than "miraculous", requiring nuclear reactions), or a resurrection from the dead is, in the case of most persons educated in a wide range of science and its ramifications, now met by disbelief. The case for an external intelligence which might temporarily have changed the laws of nature, or to whom *Homo sapiens* might be something special, finds no support from the advances in science over the last two centuries. Far from "God created Man in his own image", the evidence is tellingly summarised in "Man created God in his own image". The true builders of *Homo sapiens* are molecules, acting within the bounds of the forces and other laws that constrain them, and with their evolved methods – both benign and brutal. And "Man" also erroneously created a mind divorced from body.

Matter and Mind

...angels are spiritual creatures with no bodily component. They have intellect and will, and are much cleverer than we are.

– Christopher Howse, writer on faiths 'Daily Telegraph' London 4 September 2009

Faith-based conclusions are often very different from those based upon evidence. From evidence comes the conclusion (chapters 9 and 14) that mind, consciousness and intelligence all arise from the functioning of cells. Mind derives from the activity of matter, and intellect requires body. Changes between the physical states of the brain that allow "a state of mind" can no longer occur following death, for cellular function requires a flux of energy.

Arthur Kornberg shared the Nobel Prize in Physiology or Medicine for 1959 with Severo Ochoa "for their discovery of the mechanisms in the biological synthesis of ribonucleic acid and deoxyribonucleic acid". As a Professor of Biochemistry at Stanford University, he was also the first person to reassemble the molecular components of a virus in a test tube and, in so doing, to reassemble a "living" virus – the quotation marks necessary because a virus is not a living organism in the true sense but rather survives as a parasite of living organisms. It reproduces itself with amazing efficiency and rapidity, but is only able to do so by employing the biosynthetic apparatus of the invaded organism.

Arthur commented:[8] "Can we come as close to understanding the mind and human behaviour as we have (to understanding) metabolism? The first and formidable hurdle is acceptance without reservation that the form and function of the brain and nervous system are simply chemistry. I am astonished that otherwise intelligent and informed people, including physicians, are reluctant to believe that mind, as part of life, *is* matter and *only* matter."

It is false to speak of "mind over matter". Mind is the outcome of matter – allowing the licence to include photons – acting upon matter. When faith-based "healers" ply their trade, they bombard the receptors of the patient with some combination of photons, sound waves, changes of pressure, etc. Through the placebo effect (chapter 14), positive expectations may promote healing. "Faith", even if based on a demonstrably false assumption, can be beneficial. But where the positive expectations are constituted from a mumbo-jumbo that discourages more efficacious treatment, they often lead to disaster:

> *But Alexandra (last Tsarina of Russia) had a fatal flaw: the gift of faith. She was like the White Queen who had trained herself always to believe ten impossible things before breakfast. Once she had made up her mind that something was true, it became so. Nothing would shake her belief.*
>
> – Andrew Cook, 'To Kill Rasputin', Tempus Publishing, 2005

Since we are emotional and vulnerable creatures, it is understandable that the last Tsarina clung to the belief that the remissions of her son's haemophilia were to be credited to Rasputin's magical powers as a holy man – one who notoriously used prostitutes. And even if the Tsarina was "clutching at straws", faith could, even if based on a false premise, possibly benefit her state. So the conclusion that she had "*a fatal flaw; the gift of faith*" is not 100% secure. Positive expectations can even enhance the immune system, and may in some cases give help in fighting disease. But experience shows that serious disease is very rarely cured by positive expectations alone; science-based treatment enormously more likely to be beneficial. In the twenty-first century, Alexandra would have the option of not bearing a son with haemophilia (type A); for, on average, only fifty per cent of the male offspring of female carriers of the trait have the disease. Hence, through genetic screening, she could have borne a son who would not suffer in this way.

The heeding of Rasputin-like figures has been pushed back in some parts of the world. But immortal souls are still very much alive.

The Departure of Soul – And Other Things

> *The soul of man is immortal and imperishable.*
> – The Greek Philosopher Plato in 'The Republic'

> *Philosophers say a great deal about what is absolutely necessary for science, and it is always, so far as one can see, rather naive, and probably wrong.*
>
> – Richard Feynman

Feynman is rather hard on philosophers, for the best have posed challenging and fascinating questions. But his comment, like the one in a similar vein by Boltzmann before him, is fair about the weaker ones. Science concludes that the concept of a human soul [referring (as defined by the *Oxford Dictionary*) to "the immaterial part of man *(woman)*, eg immortality of the soul"] must go out of the window.

The problem was evident many years ago. The fossil record shows that humans evolved, relatively recently and gradually, from creatures rather similar to contemporary apes. Thus, if soul is an attribute unique to *Homo sapiens*, over what period and how did humans acquire it? Alternatively, those in favour of an immaterial and immortal soul might postulate that (like intelligence, mind and consciousness) it might be a property that emerges as organisms evolved to have increasingly complex lifestyles. But there are problems for this view also since intelligence, mind and consciousness function by courtesy of molecules. How could immaterial souls evolve from functions that require matter?

So, the concept of an immaterial soul, which could exist in isolation from bodily functions, has not stood the test of time. And neither has science found evidence for gods, angels, "spirits", ghosts, demons or the devil. Nor for places — heaven and hell — to which some believe the migration of souls occurs; nor for miracles, which require suspension of fundamental scientific laws.

A considerable fraction of the Earth's population still adheres to many of the above concepts. A major article on angels was the cover story for *Time* magazine on 27 December 1993.[9] When a sample of US citizens were asked if they "believe in the existence of angels", 69% answered "yes" and 46% believed they had their "own guardian angel". The story went on to say that 55% of the people surveyed believed angels are "higher spiritual beings created by God with special powers to act as his agents on Earth", and 49% said they believed in the existence of "fallen angels or devils". In 2005, a course that included advice on how to carry out an exorcism was held at the school of theology at Regina Apostolorum, a pontifical university in Rome. It was reported that contacts with satanic sects, by around half a million Italians, made necessary the advice; the concept of driving out demons alive in Europe in the twenty-first century.[10]

When metabolism has ceased following death, mind is gone. This is an argument against a "soul" that can survive the death of the organism; and not against an emotional experience expressed in terms of being felt "in the soul". Such emotions do not require an immaterial part of a human, which might also be given the property of immortality. The sentiment is conveyed by Peter Atkins in *The Creation*:[11] "What wonder there is (of life) should, in my view, not be at the benevolence and subtlety of external intervention, for this leads to the unnecessary intrusion of a spirit and the invention of a soul. It should instead be wonder at the realisation that underlying simplicity can have such glorious manifestations when elaborately coordinated, and that such coordination can grow through the selection of evolution."

And it is not only immortal souls that die slowly – sometimes our selves.

Euthanasia and Suicide

O, that this too, too solid flesh would melt, Thaw, and resolve itself into a dew, Or that the Everlasting had not fixed His canon 'gainst self-slaughter.

– William Shakespeare in 'Hamlet'

The definition of euthanasia used here is "bringing about, in the case of incurable and painful disease, an easier death than that which would otherwise be highly probable". Most commonly in European and American states, euthanasia is against the law and classified as a criminal act. Specifically in Britain, euthanasia remains covered by the law of murder (although as of September 2009, the situation is under review with regard to some cases).

Despite the above prohibitions, the deliberate administration of medication to facilitate death, for reasons that are perceived to be humane, is practised in many places. And by the beginning of the twenty-first century, gradual change was coming about in some parts of North America and Europe. The US state of Oregon, Belgium and the Netherlands permitted euthanasia (the first through "assisted suicide"). Switzerland does not allow doctors to practise euthanasia, but private suicide facilitation is legal. Some have travelled to Switzerland to make use of this law. But it is against Swiss law for someone to facilitate a suicide if they act out of self-interest – a situation that, on occasions, may be difficult to judge.

The changes that are occurring appear likely to be influenced by several factors: a consequence of the advanced medical technologies now available to prolong life even in cases where it is clear that a life of quality cannot be restored; wide dissemination through television of the tragic loss of faculties of some at an advanced age; increased knowledge with regard to our condition; and increasingly humane ways of aiding a painless death. The science establishing that functioning cells are required for life leads to the view that the concept of "self" has no more reality following death than prior to conception. To those holding this view, being dead – although a bleak thought – holds no fears.

But dying unsurprisingly holds strong fears, as does also the often-preceding decay of physical and mental faculties. The broad medical definition of dementia is "loss of mental ability". Loss of *some* mental abilities is often not in dispute. But these abilities are of incredible diversity, and frequently only superficially understood. So, giving proper consideration to the wishes of the patients is fraught with difficulties. The frustrations and sufferings of some patients who have lost various physical and mental capacities, but retained others, can probably not even be imagined by carers; their views, often conveyed with great difficulty, perhaps strongly opposed or not fully understood by those into whose care they have been delivered. If both parties are seriously frustrated and hampered by the different views then confrontations – sometimes violent – arise.

A common solution is then administration of anti-psychotic drugs, repulsively described as "the chemical cosh" – now judged in the UK to be a factor in the deaths of about 2,000 patients per annum.[12] A controlled study published in the British Medical Journal concluded that: "Those living in nursing homes receive poorer care than those living at home in terms of underuse of beneficial drugs, poor monitoring of chronic disease, and overuse of inappropriate or unnecessary drugs."[13] Experience shows that many nursing homes are good, but also that a revamping of the treatments administered in some of them is necessary.

The prospect, or reality, of dying slowly in a state of severe depression and/or pain naturally engenders fear. Therefore, it would seem humane to place the greatest importance upon the wishes of the individual. Since *in extremis* it may be impossible to express these wishes, in developed countries they are increasingly expressed in advance in "living wills" (which are commonly not legally binding). Individuals taking this view regard euthanasia as humane and justified in those cases where extending life involves continuing suffering that is severe, and the situation is unlikely to improve; involves an individual who, if conscious and able to communicate, has consistently expressed a preference not to waken from the next sleep. Under such circumstances, the insistence of the rest of society to impose the preservation of life of the sufferer seems cruel – the commonplace "not something that considerate dog owners impose upon their pets".

Yet, consistent with the most common state of the law, the extension of such a state of severe suffering receives support. An article under the headline "The old and the sick shouldn't be given a quick exit button" appeared in a UK broadsheet newspaper in 2005.[14] Opinion of an emeritus professor of surgery that "the wishes of patients have never been an overriding ethical imperative for doctors" was cited. The Christian Medical Fellowship opposed a Bill proposing "Assisted dying for the terminally ill".[15] Their website reports faith leaders getting together to oppose such assistance. Views that "the right to life of the old" are eroded are correctly expressed, but the wish that the old sometimes fervently wish to surrender that right is ignored. Euthanasia splits communities with strongly opposing reactions, comparable to those expressed on the issue of abortion – with its extreme propagandas of "pro-life" and "pro-choice".

In addition to the widespread opposition to euthanasia, almost all major groups and belief systems discourage suicide – to use Shakespeare's phrase: *"self-slaughter"*. The extent to which suicide is ruled to be a criminal act has decreased in modern times. Suicide is not a crime in England and Wales, having been decriminalized in 1961. But there is a possibility of up to fourteen years imprisonment for anybody assisting a suicide; it is a crime to assist in a non-crime.[16] Unusually, Germany has had no penalty for either suicide or assisted suicide since 1751.[16]

Opposition to the suicide, assisted or otherwise, of someone suffering from Huntington's chorea, a disease accompanied by highly distressing mental deterioration in middle age, and invariably fatal, is contrary to the humane instinct of many. The disease commences with personality disturbance and involuntary movements, and makes an inevitable progression to death. It is caused by long repeats of the DNA triplet CAG in the long arm of chromosome 4. Through the genetic code, these repeats are translated to a protein that has a long tract of repeated glutamic acids. The balance between tragedy and normality is a fine one. If the number of CAG repeats in the relevant gene, which we all possess, is less than 27, there is no disease; if greater than 39, a tragic fate is sealed through the genetic lottery.

But the "thin end of the wedge" argument surfaces frequently. For the inhuman side of humanity is expressed when some wish to benefit from the death of others – presumably a factor which relates to the law being more considerate to dying dogs than to humans. Additionally, some religious groups express opposition because "life is sacred" ("held dear to a deity"). "Life is sacred" seems a dubious clarion call in a world where no mammal can survive without eating another life form; where a large fraction of humans are omnivores, with at least ten billion chickens killed and consumed per annum; where 1.5 million young lives are lost each year due to a lack of the most basic sanitation[17] – sanitation which could in principle readily be supplied by affluent systems.

In few places is "Why Freedom Dies" (chapter 12) more evident than the manner of dying. The exercise of more humane behaviours falls victim to dangers both real and imagined, to emotions and to dogma. Newer knowledge points not to "life is sacred"; rather to the conclusion that "the frail must be treated humanely" – not least in the closing stages of life. This newer knowledge also points to a desirable policy: improve environments and educate in deductive reasoning on a global basis. Whether such a policy will evolve and, more challengingly, be brought to fruition is a big question.

Other "Big Questions" sometimes attract a media frenzy, as when the mathematical physicist Stephen Hawking in 2010 claimed, "The universe can and will create itself from nothing."[18] When mathematical approaches to science lead to great claims, experimentalists will demand evidence (as spectacularly provided for Einstein's $E = mc^2$). Hawking's claim currently lacks such evidence, and wise heads will defer judgement. But "Big Questions" do merit more reflection, and are raised in closing this account of the journey of matter.

17

Big Questions

Ultimate Questions

Arguments based on good design, natural beauty, perfection and complexity of the world or universe are a major reason why people believe in God.

– Conclusion from Michael Shermer, Intellectual and Emotional
Reasons to Believe, based on his book 'Why We Believe', 1999

In the foregoing chapters, I have avoided big questions where our ignorance precludes reasonably secure answers; or where the reasoning is so abstruse that it is difficult to convey to the generalist. For example, the entropy of black holes; the origin of matter; the concept of an unimaginably dense singularity; why mathematics has often led to excellent theories of the behaviour of matter; the profession of a clear understanding of what constitutes consciousness. But current knowledge does throw clear light on other "Big Questions".

When Charles Darwin asked the Big Question with regard to species, he looked at a wide spectrum of organisms and reached his revolutionary conclusion. He had passed through the Universities of Edinburgh and Cambridge and was well primed. In Cambridge, he had met Adam Sedgwick, the professor of geology, and John Henslow, the professor of botany. Both these men were addressing big issues at the then frontiers of knowledge. But he was only "primed", and no more; for Sedgwick was a deeply religious man and one later grieved by Darwin's eventual conclusions.

After Cambridge, Darwin embarked on his epoch-making voyage on HMS Beagle, bringing a mind – one that questioned and reasoned – into a wide range of new environments. Long before he reached South America and the Galapagos, he visited the Cape Verde Islands. The vision was "like giving a blind man eyes – he is overwhelmed with what he sees and cannot justly comprehend it".[1] On reaching Tierra del Fuego, he observed the local human population, surrounded by hostile tribes, and observed, "The cause of their warfare would appear to be the means of subsistence."[1] Worse was to come for the Fuegians, for migrants to the New World actively sought their extermination, and none of pure blood are alive today.

He saw that life in many places is inevitably very tough; in some of its forms, emotionally repulsive. In Chile, upon observing the turkey buzzard, he commented, "This disgusting bird, with its bald scarlet head formed to wallow in putridity..." Also while in Chile, an interest in barnacles was stimulated on observing a species of parasite, which he named *Cryptophialus minutus*, that bored through barnacle shells to reach the inside of the unfortunate host. In these parasites, the males are less than one hundredth of the mass of the females and lack all organs except for testicles, seminal vesicles and "a wonderfully elongated prosbosciformed penis".[1] Such was the effect of these careful observations that a decade later at his home in Down House, he did serious taxonomic work on barnacles – to the extent that one of his children asked a friend, "Has your father done his barnacles yet today?" Barnacles, never even seen by a large fraction of the Earth's population, and studied in detail by almost no one, were part of the story that led to the rejection of grand design.

One hundred and fifty years after the publication of *The Origin of Species*, some physicists of religious persuasion – few in number – produce books in support the idea of an external intelligence, to which the evolution of humans is a purpose. What impresses these authors in reaching their conclusions? First, they point to the laws of physics which can be expressed in precise mathematical form; and to the small number of fundamental forces. They ask, "Why are the rules so simple, precise and beautiful?" Second, they note that the initial conditions in the "Big Bang", and the relative masses and properties of the limited number of fundamental particles, are such as to have given a Universe just right for the formation of stars and planets. If the rules were different in even a remarkably small way then neither the Universe nor humans would exist. They are struck by the fact that it appears that it could have happened in countless billions of other ways – but it did not. As an answer to the question "Who devised these conditions and rules?", they say, "Let's call it God."

The above thinking is anthropomorphic; God with *Homo sapiens* in mind; the assumption of a "who" an enormous leap of faith. But the case against a loving and omnipotent creator is not found in physics. It is found in biology. For the rules give rise to organisms with transparently unloving traits; to vultures that spend part of their lives literally "down to their necks" in blood; to the agents of syphilis and leprosy. The rules lead to a globe in which roughly one half of the current human population ekes out an existence on less than \$2 a day – with all the inevitably sad consequences.

So, we have two very different conclusions with regard to one of the "Big Questions": a benign creator God with a purpose; or the Darwinian view – given rules, selection of whatever serves survival. In addressing these differences, three "ultimate questions" are posed, following in the footsteps of the scientific philosopher Karl Popper and the distinguished medical scientist Peter Medawar.[2] In broad terms, they ask, "What is the purpose of life?"; "How did everything begin?"; "Why are there precise rules that control the behaviour of the Universe?"

Does Life Have a Purpose?

In his book *The Mind of God*,[3] the physicist Paul Davies concludes:

> *I cannot believe that our existence in this Universe is a mere quirk of fate, an accident of history...the existence of mind on some planet in the Universe is surely a fact of some significance. Through conscious beings, the Universe has generated self-awareness. This can be no trivial detail, no minor by-product of mindless, purposeless forces. We are truly meant to be here.*

Human self-awareness and consciousness are often used in propounding theories of God and "human purpose". But *all* creatures are made up of molecules, and it is activity within molecular structures that gives rise to self-awareness – a property that gradually emerges, and does not appear at the drop of a hat. These properties, in their emerging forms, appear likely to be part of the experiences of our mammalian relatives. Chimpanzees plan ahead and a mother chimpanzee will grieve upon the loss of her baby; dogs appear to dream, and also to plan ahead. If my wife and I pack bags, our dog immediately takes a position next to the car to ensure that he will not get left behind. If the table is set in the dining room, as opposed to the kitchen where we more frequently eat, he concludes "visitors soon likely to arrive". He then takes up a position close to the entrance to the house, his eyes fixed on the door. Elephants gently caress the bones of their dead with their trunks and feet while behaving in a solemn manner. They appear likely to have an awareness of the death that befell their ancestors.

The conclusions of Paul Davies are bold – forces have a purpose, and the purpose is to produce the conscious beings *Homo sapiens*. The question of "purpose" brings us to the subject of teleology – the notion that organisms were designed for some "purpose". Jacques Monod[4] pointed out that almost all the important ideologies of the past, and many of the present, are based on a belief in teleology – a sense of purpose. While it is convenient to lapse into teleological language – say, "the purpose of the kidney is to remove metabolic waste from the body" – the idea of natural selection removes the idea of purpose.[5,6] Complex systems evolve to produce sophisticated *function*: there is no evidence for purpose or for design.

The *sense* of purpose in a life derives from the requirement of successful function. The requirement simply of function is a view that most people readily accept if applied to a virus; a bacterium living in the human gut; or perhaps even to a spider. However, as the evolutionary timescale progresses to creatures with brains as complex as those of dogs, chimpanzees or humans, human societies increasingly speak of "purpose". What is seen is actually increasingly sophisticated *function*.

How Did Everything Begin?

The apparent requirement of "a beginning" may be an artefact of the workings of our brains – an organ that has evolved with experience only of systems that begin and end. So, the question itself may make an invalid assumption. For the Universe could have an infinite past, given that one scenario is that the Big Bang commenced from the collapse of a previous state. However, that the present Universe may eventually collapse inwards on itself from the present expansion currently seems improbable in the light of the dark force – the repulsive force that opposes gravitational attraction.

But all we know is that there was at least one "Big Bang", about 13.7 billion years ago. "Why there is something rather than nothing" currently lacks a clear answer. But the scientific method demands we do not throw up our hands; do not appeal to mysticism. Rather, regard the problem as a future challenge. "God the Creator" provides no understanding for, as seen earlier (chapter 16), such an argument involves an infinite regression.

Why are there Rules?

> *If there is a God that has special plans for humans then He has taken very great pains to hide His concern for us.*
>
> – American physicist Steven Weinberg in 'Dreams of a Final Theory'

Rules are necessary for the existence of the Universe and its diverse components. But only four (or possibly five) forces, in conjunction with a remarkably limited number of laws, appear sufficient to describe broadly (but not more) the macroscopic Universe, organisms and the molecules they contain. But we do not understand why the forces are of the nature that they are; or have the magnitudes that they do. Nor do we possess a theory as to why the "fundamental" particles are of the number that they are; nor why they have the properties that they do. The "grand unified theory", much desired by physicists, is notable for its absence.

But Darwin's ideas – of variation, fitness and selection – give a wonderfully unified view of what has happened throughout the journey of matter. Even outside biology, these principles guide understanding: the evolution of the celestial bodies, the evolution of cars and of clocks. As in the evolution of organisms, new combinations and interactions produce new things, the "fittest" of which is selected. The idea that variation, fitness and selection apply to the evolution of the celestial bodies is an unconventional one. But heterogeneity in an expanding Universe supplied variation. In some environments, having the largest mass conferred the greatest fitness, for black holes survived very well by "eating" their neighbours. In other environments, a mass that switched on fusion but at a slow "metabolic rate" conferred fitness relative to bigger stars that soon burnt out. Here, selection favoured smaller citizens. Size matters, and is selected according to environment.

Perhaps variation, fitness and selection provided a Universe that, although seeming incredibly improbable (chapter 2), was furnished in the same manner as other seemingly improbable systems – flies, spiders and humans. It may be speculated that the original forces and particles were diverse; and that from this diversity, the forces and aggregates of particles that later prevailed were selected precisely because they produced a Universe and constituent parts that were stable. Although there is no evidence that, at an early point in the Big Bang, the four forces were selected to satisfy this requirement, the aggregates of particles that later successively come into being *were* selected in this way. For throughout evolution, particles combined to produce many new entities (variation) that then allowed the selection of those aggregates that had good survival properties in the prevailing environments. A different speculation to get over the probability problem – the existence of a large number of parallel universes ("multiverses") – lacks this attractive feature.

In a 2009 interview for the *UK Radio Times*, David Attenborough commented on hate mail received from TV viewers for failing to credit God in his documentaries:

> *They always mean beautiful things like hummingbirds. I always reply by saying that I think of a little child in East Africa with a worm burrowing through his eyeball. The worm cannot live in any other way, except by burrowing through eyeballs. I find that hard to reconcile with the notion of a divine and benevolent creator...Evolution is not a theory; it is a fact.*

Plant cells grow; a caterpillar eats the cells; the caterpillar is eaten by the offspring of a parasitic wasp; a small bird then eats the offspring; the small bird is then eaten by a hawk; the hawk dies and is eaten by carrion; when the carrion species dies, the molecules from its body – transformed by bacteria, fungi and myriad forms of small animals – populate the surrounding soil. A plant uses these molecules, and so the cycles continue. Given information on these cycles, millions of times more efficient in their recycling than current human societies, individuals can conclude whether they, and their fellow creatures, are the product of mindless forces or of a benevolent intelligence.

That female spiders of the species *Latrodectus mactans* sometimes eat the male after copulation has been known since the 1930s, causing them to be called black widows. But in the case of *L hasselti*, otherwise known as the Australian Redback spider, the male is even more likely to be eaten apparently because he asks for it. To cite Lyn M Foster of the University of Otago:[7] "After the male inserts one of his two sexual organs (or emboli) into the female, he back flips into her jaws. Copulation proceeds while she slowly masticates his abdomen and injects enzymes. At the end of a possible second copulation, the male is already half digested, whereupon the female wraps him in silk and concludes her repast." Those speaking against the maxim "God is love" feel churlish but are entitled to note that, if He exists, His creatures are remarkably diverse in their tastes.

Scientific investigation has informed us that forces and laws that know neither good nor evil controlled the journey of matter. As Bill Bryson put it, "Life just wants to be."[8] The scientific method points to the conclusion that survival is served by the human pro-social qualities of goodness, love, loyalty and cooperation on the one hand; but both served and sullied by competition on the other.

Flaubert noted that selfishness is a necessary condition for human happiness. For about 300,000 humans die every day; yet, for day-to-day happiness, it must be possible to soon put aside the great suffering associated with many of these daily deaths. Among many, Charles Dickens, Honoré de Balzac, Victor Hugo, Franz Kafka and George Orwell expressed in their novels the inhumane side of humanity; among many, Epicurus, Spinoza, Pierre-Simon Laplace, John Stuart Mill and David Hume promoted rationality and liberal views. They, and their many like-minded authors and thinkers, might have valued the recent discoveries that indicate the molecular origins of both the appealing and appalling parts of human behaviour.

Individuals, institutions and nations typically ensure and insure their survival to a degree that makes charity a restricted currency. Exceptions are found when they are so rich that their security lies beyond reasonable doubt – as in the case of thirty-eight US billionaires who, in 2010, pledged to give half of their wealth to charity either during their lives or after their deaths. A survey of the web indicates that, at the national level, the relatively liberal Scandinavian countries are the most generous in giving foreign aid, at about 1% of GDP. Most other affluent nations compare poorly with this figure; often with some of their charity thrown into an ambiguous light – more aid given to regions where their political interests are best served. With regard to charity that is internal, as opposed to international, the USA has a relatively high figure (about 2% of GDP); possible influences include national affluence and religious conviction, coupled with efficient sales pitches.

When the news breaking on a single day (1 December 2007) includes the following items,[9] it appears that reaching agreement on what is humane and rational will be, at best, difficult. First, reports appeared of Pope Benedict XVI's second encyclical. It stated that atheism "has led to the greatest forms of cruelty and violations of justice".[10] The encyclical might have in mind the barbarous regimes of Pol Pot, Stalin and Mao – highly appropriate. However, cruelty has been common at some point in almost all institutions and cultures. It would not be expected that the Catholic Church would be exceptional in this regard – and it is not (chapter 13). The record establishes that some Christians, like some atheists, have developed problematical, or even brutal, behaviour. Others of each persuasion have epitomised gentility, and it would be wonderful if *all* these gentle people could be praised. A blanket association of atheism with cruelty and violations of justice opposes consensus.

The complexity of the situation is illustrated by the Catholic Church's current opposition to contraception for Africans in a continent under the scourge of AIDS – a sexually transmitted disease which, in 2005 alone, killed an estimated two million people in sub-Saharan Africa.[11] The Church has a moral precept that intercourse must allow the possibility of pregnancy. It is correct in arguing that abstinence from sex – as an alternative to contraceptive-protected sex – must lead to a decline in AIDS. But the realities are that abstinence from sexual lust by a large body of males has not proved enforceable by decree. This truth is witnessed by the past behaviour of a significant percentage of Catholic priests, the requirement for celibacy leading males into socially unacceptable behaviour. Pragmatists therefore argue that humanitarian action to restrict AIDS in Africa requires the provision of male contraceptives; argue that justice is thereby more probably available to the newborn child who otherwise is often already infected and condemned to suffering and early death.

The encyclical goes on to proclaim: *Loving God requires an interior freedom from all possessions and all material goods.* This statement sits oddly with the material wealth and ostentatious display of the Vatican, its "material goods" contrasting sharply with the poverty of many of the adherents to its faith. But material wealth and ostentatious display go hand in hand with influence and power. Realities are served for, in the contest to survive, institutions make attempts to perpetuate their binding beliefs and power – even if these involve selective amnesia and blinkered observation.

A second report on 1 December 2007 was of a Harris Poll of 2,455 US adults conducted online by Harris Interactive between 7 and 13 November 2007.[12] According to the report, a higher percentage believed in the devil than in Darwin's theory of evolution (62% *vs* 42%). At the outset of the twenty-first century, there was no significant move towards evidence-based belief in the USA, for the survey concluded that, in comparison with a 2005 poll, the fraction believing in miracles, angels and witches had slightly increased. A 2006 survey of the acceptance, or rejection, of evolution reported[13] that although 65 to 85% of those living in the western part of Europe, or Scandinavia, accept it as true, this figure was again low (39%) in the USA, with an almost equal fraction believing the theory to be false (around 22% were "not sure"). Among the nations in the 2006 survey, it was concluded that "Only adults in Turkey...were less likely to accept the concept of evolution than American adults."

Because of conformity to conventional wisdoms, it is possible that there is inflation of the majority figures for the above beliefs. But, clearly, the evidence–based story of science, as told in these pages, has not penetrated widely. The story admits that there is much left to learn. For dark energy and dark matter – about which we understand almost nothing – account for about 95% of the mass/energy of the Universe; development is only poorly understood; consciousness has yet to reveal its secrets. But the story tells that environments and biology control what we are; and that these environments and the molecules of biology are in turn crucially fashioned by the four very different forces – strong and weak nuclear, electromagnetic and gravitational.

The story tells that we understand how order arises spontaneously from chaos. The story tells of the universality of atomic nuclei, the elements and of the diversity of molecules that they make; of living systems that are possible without design, each with its common features of DNA, membranes and proteins. It tells of the universality of genes, and of the genetic code that we share with flies and spiders, and demonstrates a universal synthesis from only protons, neutrons and electrons to make all that matter we know. It pronounces on the widespread beliefs in benign gods as probable falsehoods;[14,15] and on the probable origins, and sometimes bizarre nature, of these beliefs.[16]

It is the information deriving from physics, chemistry and biology, and the Darwinian synthesis that leads to a more enlightened understanding.[17,18] It is through experiment, and an understanding of what has gone before, that we know that the ape brain is only partially over-written; that free will is a delusion. Pinpointing this falsehood allows a clear understanding of why millions, who might be saved, are left to die. The controlling laws are so powerful that even our understanding of them has not yet permitted the forging of the desired humanity. The blind forces work their principles, whether acting upon extraordinarily complex animate, or relatively simple inanimate, systems. Those rejecting blind forces must provide a better model to account for the loss each year of those 1.5 million young lives due to a lack of the most basic sanitation[19] amidst the opulence of powerful systems.

Despite some gloomy facts, there is an encouraging trend evident in biology – with the passage of time, not only the evolution of more efficient ways of out-competing some but also, in the interests of survival, increased cooperation with others. As time has passed, individuals, and the molecules and cells within them, have increased their cooperation as a consequence of genetic selection. In this way, extended systems with increased stabilities were forged (chapter 11).

Increases in human cooperation due to changes in genes are likely to be at a snail's pace compared to the hare of current cultural change. But behavioural change through changes in gene expression has been greatly accelerated in our mammalian cousin the silver fox. The acceleration was achieved by intense selection pressure.[20] Starting with a population of wild foxes, it was found that, after one generation, a few per cent of the offspring showed less aggressive behaviour. These variants were selected as breeding pairs, and a similar intense selection process was followed at each succeeding generation. By only *the tenth generation*, friendly "dog-like" behaviour had been selected in about 75% of the offspring – sharply in contrast to the usually viciously aggressive behaviour found in the starting population.[20,21]

This is an important result, demonstrating that, under intense selection pressure, a close mammalian relative undergoes a rapid change in genetically coded behaviour. There is also confirmation of a trend already noted by Charles Darwin: concurrent with the rapid evolution of their socially cooperative behaviour, the foxes increasingly retained juvenile features (eg floppy ears and broader skulls). It appears that some of the genes that promote aggressive behaviour are master genes, for they also control the expression of genes that lead to adult characteristics. This is not too surprising for, during the evolution of carnivores, survival required plenty of aggressive behaviour in adults. The correlation simply represents efficiency in genetic programming – adult features and aggression under the same controlling switches. The retention of juvenile features in socially cooperative pets increases their attractiveness to humans – they become surrogate children.

In the words of the senior Russian author of this study, Lyudmila Trut: "Before our eyes, 'the Beast' has turned into 'Beauty'."[21] So, a human mind had mitigated a genetically coded aggression consequent upon the mindless forces. That mitigation was possible because the mindless forces had produced a brain that could store and process information, and that brain experienced an environment that provided the appropriate knowledge – the ultimate super-computer in a so far little-occupied niche. Since about 75% of known human genes have an equivalent in dogs, an analogous programme of intense selection in humans would likely lead to changes in the same direction in our species.

But conscious, intense genetic selection against aggressive human behaviour seems unlikely in the extreme. For, although the aim would be to reduce violent traits, to produce an outcome opposite to that desired by unsavoury fascist dictators, the practice is still that of eugenics. The evidence points to severe dangers. Residual aggressors would likely exploit the evolving meek and mild. And, in selecting for a desired trait, master switches may cause the concurrent appearance of much less welcome physiology and behaviour – witness the problems of highly inbred dogs.

Although conscious selection for pro-social behaviour is anathema, it has been going on at an unconscious level for centuries as human population densities have increased. Parts of the global system are being increasingly directed on this path through highly transmissible culture. Retentive brains, fed by the worldwide web and global television, may direct our species to greater cooperation. In this increasing cooperation, the more an entity can bring to the table, the more likely it is to be absorbed as a cooperative participant. This much is evident when couched in the language of the physical and biological sciences. In the process of crystallisation, impurities – unable to cooperate in the process – are squeezed out of the cooperative lattice (chapter 11). The plight of much of central Africa bears analogy to this process. This part of Africa currently comes to the table as a poor relation and struggles to reap benefits from globalisation that outweigh the costs of its exploitation. In general, the poorest nations tend to be squeezed out of the cooperative system. Through negative feedback, their ability to provide a global service declines as does, in parallel, the extent to which others wish to cooperate with them. They suffer a neglect that is tragic.

The televised spectacle of the largest collection of predators – a mix of mammals, fish and birds – on Earth frenziedly eating their way through in excess of a hundred million sardines[22] diverts some of our species away from the concept of a loving creator; a concept simply inconsistent with this spectacle, and the wealth of evidence cited in these pages. Through a more global knowledge of the ways of Nature, there is a clarification of the roots of problems. The horrors of conventional solutions are demonstrated; more sophisticated and humane solutions may be sought.

The lines of Oscar Hammerstein II in his musical *Showboat* encapsulate the enormity of the problem of dealing with deeply programmed behaviour: "Fish got to swim, birds got to fly, I got to love one man till I die, Can't help lovin' dat man of mine." Fish, birds, humans – the survival machines constituted from endless dances of sub-atomic particles connected in webs of Byzantine complexity. So much so that when a fortunate sees another human in pitiable condition, the relevant reflection is: "There, but for the grace of my genetic make-up and environment, go I." Those who reject this conclusion must come up with a better understanding of the world system. A system in which UNICEF reported for 2006 a death rate of almost ten million children (below five years of age) per annum.[23] These deaths – which still continue at a rate of around 20,000 per day – are largely due to malnutrition, lack of a clean water supply, or diseases readily cured in wealthy countries. These are all problems that, in principle, can be tackled; and all were occurring as ski slopes were built in Dubai.

In the Head's report on humanity, "Has obeyed the rules" is written on the basis of the evidence. Dissent from this conclusion is inevitable, for "Has obeyed the rules" flies in the face of much received wisdom; is too bleak to countenance. Some believe in choice that is free, whatever the experiences of the individual. This view has not solved the tragedy of the ten million children. Others simply accept human nature as it is; this cannot solve their tragedy. It is globally improved environments that could promise a more benign future (chapter 16). They can become more probable through the widespread dissemination of the new knowledge. They hold out the hope that, in future, the Head's report might not only say "Has obeyed the rules" but also "Has done better".

The manner in which humanity was forged poses a challenge to efforts to forge greater humanity. But reprehensible behaviour, derived from genetic memory and bleak environments, may be mitigated even over a period of only a few generations through improved environments. Through any book, something new comes into our environment and, whether the new is little or large, some behaviour is modified. If these pages promote a message of hope without recourse to the fictions that are less taxing than truths, an intended function will be served.

References and Further Reading

Chapter 0: Truths and Concepts

1. Karl Popper, *The Logic of Scientific Discovery* (translation of *Logik der Forschung*) (Hutchinson, London, 1959).
2. Steven D Levitt and Stephen J Dubner, *Freakonomics: A Rogue Economist Explores the Hidden Side of Everything* (Allen Lane, Great Britain, 2005).
3. C Djerassi, *Calculus* – a play on the priority in the discovery of calculus, which documents the bitter dispute between Newton and Leibniz (2004).
4. G Gamow, *Mr Tompkins in Paperback* (Cambridge University Press, Cambridge, UK, 1965).
5. See http://imagine.gsfc.nasa.gov/docs/science/mysteries_l1/age.html
6 M Rees, *Before the Beginning* (Simon and Schuster, London, 1997).
7. J J Halliwell, *Quantum Cosmology and the Creation of the Universe* (Scientific American, December, 1991).
8. S Hawking, *A Brief History of Time* (Bantam Books, London, 1988).
9. D Goldsmith, *The Astronomers* (St Martin's Press, New York, NY, USA, 1991).

Chapter 1: A Cooling Universe

1. S Weinberg, *The First Three Minutes* (Andre Deutsch, London, 1977).
2. B Mahon, *The Man Who Changed Everything – the Life of James Clerk Maxwell* (J Wiley, Chichester, UK, 2003).
3. M Rees, *The Anthropic Universe* (New Scientist, 6 August, 1987).
4. R P Feynman, R B Leighton and M Sands, *The Feynman Lectures on Physics* (Addison Wesley Publishing Co, Inc, Reading, Mass, USA, 1964).
5. B Greene, *The Elegant Universe* (Jonathan Cape, London, 1999).
6. S Woosley and T Weaver, *The Great Supernova of 1987* (Scientific American, August, 1989).
7. W Heisenberg, *Physics and Philosophy – The Revolution in Modern Science* (Harper, New York, 1958).

Chapter 2: From Stars to Living Room

1. M Rees, *Before the Beginning* (Simon and Schuster, London, 1997).
2. www.strangescience.net/hutton.htm

3. L Pauling, *The Nature of the Chemical Bond and the Structure of Molecules and Crystals: An Introduction to Modern Structural Chemistry* 3rd Edition (Cornell University Press, Ithaca, NY, 1960).
4. M Rees, *The Anthropic Universe* (New Scientist, 6 August 1987).
5. J Gribbin and M Rees, *Cosmic Coincidences* (Black Swan Books, London, 1991).
6. J D Barrow and F J Tipler, *The Anthropic Cosmological Principle* (Oxford University Press, 1986).

Chapter 3: The Rules for Spontaneous Order

1. P Coveney and R Highfield, *The Arrow of Time* (W H Allen, London, 1990).
2. I Prigogine and I Stengers, *Order out of Chaos* (W Heinemann Ltd, 1984).
3. J Monod, *Chance and Necessity* (Knopf, New York, 1971).
4. P W Atkins, *Physical Chemistry* (Oxford University Press, 1978).
5. J Keeler and P Wothers, *Why Chemical Reactions Happen* (Oxford University Press, 2003).
6. D A Warner and R Shine, *The Adaptive Significance of Temperature-Dependent Sex Determination in a Reptile* (Nature, vol 541, pp 566-568, 2008).

Chapter 4: The Molecules of Life

1. B Alberts et al, *Molecular Biology of the Cell* (Garland Publishing, 1983; 4th edition, 2002).
2. L Stryer, *Biochemistry* 5th Edition (W H Freeman and Co, San Francisco, 2005).
3. S L Miller and L E Orgel, *The Origins of Life on the Earth* (Prentice Hall, 1974); see also A G Cairns Smith, Genetic Takeover and the Mineral Origins of Life (1982).
4. J D Watson, *The Double Helix* (Atheneum, New York, NY, 1968).
5. B Maddox, *The Dark Lady of DNA* (HarperCollins, 2002).
6. R M May, *How many Species are there on Earth?* (Science, vol 247, pp 1441-49, 1988).
7. D Bray and D. Williams, *How the 'Melting' and 'Freezing' of Proteins may be used in Cell Signalling* (ACS Chem Biol, vol 3 , pp 89-91, 2008).
8. http://www.coolest-gadgets.com/20071217/heart-hand-warmer-to-use-again-and-again/
9. See, for example, http://www.newscientist.com/article/

mg19426086.000-junk-dna-makes-compulsive-reading.html

10. C Lyell, *Principles of Geology Volumes 1-3* (John Murray, Albermarle Street, London, 1830).

11. C Lyell, *Principles of Geology an abridged version* by J A Secord (Penguin Classics, 1997).

12. A Koestler, *The Case of the Midwife Toad* (Random House, 1971).

13. R A Waterland and R A Jirtle, *Transposable elements: targets for early nutritional effects on epigenetic gene regulation* (Mol Cell Biol pp 5293-5300, 2003).

14. V Chandler, *Paramutation: from maize to mice* (Cell 128, pp 641-645, 2007).

15. L H Lumey, *Decreased birth-weights in infants after maternal in utero exposure to the Dutch famine of 1944-1945* (Paediatr Perinat Ep, pp 240-253, 1992).

16. E Avital and E Jablonka, *Animal Traditions: Behavioural Inheritance in Evolution* (Cambridge University Press, UK, 2000).

17. E O Wilson, *Consilience – the Unity of Knowledge* (Little, Brown and Co, UK, 1998).

Chapter 5: Darwin's Agents

1. C Darwin, *The Origin of Species* (London, John Murray) (A Facsimile of the First Edition Harvard Univ Press, Cambridge, Mass, 1964).

2. Janet Browne, *Charles Darwin: Voyaging* (Knopf, 1995 (vol 1); also Charles Darwin: The Power of Place (vol 2).

3. A Desmond and J Moore, *Darwin* (Penguin, London,1992).

4. E Sober, *The Nature of Selection: Evolutionary Theory in Philosophical Focus* (MIT Press, Cambridge, Mass, USA, 1984).

5. L Stryer, *Biochemistry 5th Edition* (W H Freeman and Co, San Francisco, 2005).

6. J Kendrew, *The Encyclopedia of Molecular Biology* (Blackwell Science Ltd, Oxford, UK, 1994).

7. J T Fraser, *Time, the Familiar Stranger* (J Dekker & Sons, USA, 1987).

8. J Mitteldorf, *Death by Design* (http://www.evoyage.com/JoshMitteldorf.htm).

9. R A Fisher, *The Genetical Theory of Natural Selection* (Clarendon Press, Oxford, 1930).

10. C Kenyon et al, *C. elegans mutant that lives twice as long as wild type* (Nature, pp 366, 461–464, 1993).

11. http://www.ncbi.nlm.nih.gov/bookshelf/br.fcgi?book=gene&part=hgps

12. B Best, *Mechanism of Aging* (http://www.benbest.com/lifeext/aging.html).

13. M R Rose, *Evolutionary Biology of Aging* (Oxford University Press, 1991).

14. E Lax, *The Mould in Dr Florey's Coat* (Little, Brown, London, 2004).

15. D H Williams and B Bardsley, *The Vancomycin Group of Antibiotics and the Fight against Resistant Bacteria* (Angewandte Chemie, vol 38, p 1173, 1999).

16. M R Haussler, R H Wasserman, T A McCain, M Peterlik, K M Bursac and M R Hughes, (Life Sci, vol 18, p 1049, 1976).

17. D H Williams, M J Stone, P R Hauck and S K Rahman, (J Nat Prod, vol 32, p 1189, 1989).

18. R Dawkins, *The Ancestor's Tale – A Pilgrimage to the Dawn to Life* (Weidenfeld and Nicolson, 2004).

19. E A Gladyshev, M Meselson and I R Arkhipova, *Massive Horizontal Gene Transfer in Bdelloid Rotifers* (Science, 320, p 1210, 2008).

20. For a look at Darwin's *The Origin of Species* in the light of what has followed, see S Jones, *Almost Like a Whale* (Black Swan Edition, 2001).

Chapter 6: From Bits to Bodies

1. D Attenborough, *Life on Earth* (Collins, London, 1979).

2. R Dawkins, *The Ancestor's Tale – A Pilgrimage to the Dawn to Life* (Weidenfeld and Nicolson, 2004).

3. B Shorrocks, *The Genesis of Diversity* (Hodder and Stoughton, Sevenoaks, Kent, UK, 1978); for more recent estimates, see E O Wilson, *The Diversity of Life* (W W Norton & Co, 1999).

4. D E G Briggs and P R Crowther (Eds) *Palaeobiology II* (Blackwell Publishing, p 214, 2001).

5. I Tattersall, *The Human Odyssey: Four Million Years of Human Evolution* (Prentice –Hall, New York, 1993).

6. R Dawkins, *River out of Eden: A Darwinian View of Life* (HarperCollins, London, 1995).

7. S Kierkegaard, *Repetition (Kierkegaard's Writings)* vol 6, (Princeton University Press 1983).

8. J Diamond, *Guns, Germs and Steel* (Jonathan Cape, London, 1997).

9. J Diamond, *Collapse* (Viking Penguin, 2005).

10. C Barnes, V Plagnol, T Fitzgerald, R Redon, J Marchini, D Clayton and M Hurles, (Nature Genetics vol 40, pp 1245-52, 2008).

Chapter 7: Making Bodies from Bits

1. W J Larsen, *Human Embryology* (Churchill Livingstone, 4th Edition, 2008).
2. L Wolpert, *The Triumph of the Embryo* (Oxford University Press, Oxford, 1971); see also L Wolpert et al, *Principles of Development* (Current Biology Ltd and Oxford University Press, 1998).
3. D'Arcy Thompson, *On Growth and Form* (Cambridge University Press, 1961).
4. P A Lawrence, *The Making of a Fly* (Blackwell Science, 1992).
5. For a discussion, see S J Gould, *Ontogeny and Phylogeny* (Belknap Press of Harvard University Press, Cambridge, Mass, 1977).

Chapter 8: The Needs and Forms of Bodies

1. L Wolpert, *The Triumph of the Embryo* (Oxford University Press, Oxford, 1971).
2. J Sulston and G Ferry, *The Common Thread: a Story of Science, Politics, Ethics and the Human Genome* (Bantam Press, 2002).
3. Y Gilad, A Oshlack, G K Smyth, T P Speed and K P White, *Expression profiling in primates reveals a rapid evolution of human transcription factors* (Nature, 440, pp 242-245, 2006).
4. P Khaitovich, I Hellmann, W Enard, K Nowick, M Leinweber, H Franz, G Weiss, M Lachmann and S Pääbo, *Parallel patterns of evolution in the genomes and transcriptomes of humans and chimpanzees* (Science, 309, pp 1850–1854, 2005).
5. A Naef, (Naturwiss, vol 14, pp 445-452, 1926).
6. S J Gould, *Ontogeny and Phylogeny* (Belknap Press of Harvard University Press, Cambridge, Mass, 1977).
7. S J Gould, *Wonderful Life* (Penguin Books, 1991).
8. S Conway Morris, *Life's Solution* (Cambridge University Press, 2003).
9. A Fersht, *Structure and Mechanism in Protein Science – a Guide to Enzyme Catalysis and Protein Folding* (W H Freeman and Co, 1999).
10. http://evolutionarynovelty.blogspot.com/2008/07/box-jellies-and-red-herring-of-eye.html

Chapter 9: Running and Manipulating the Human Machine

1. For details, see B Alberts et al, *Molecular Biology of the Cell* (Garland Publishing, 4th edition, 2002).
2. http://www.sciencemuseum.org.uk/broughttolife/themes/war.aspx

3. I M Roitt, *Essential Immunology* (Blackwell Scientific, Oxford, UK, 9th edition, 1997).

4. O Sacks, *The Man who Mistook his Wife for a Hat* (Duckworth and Co Ltd, 1985).

5. E R Kandel, J H Schwartz and T M Jessel (eds), *Principles of Neural Science* (McGraw Hill, New York, NY, 2000 4th Edition); see also D L Alkon, *Memory Storage and Neural Systems* (Scientific American, July, 1989).

6. W M Jenkins, M M Merzenich, M T Ochs, T Allard and E Guicrobles, (J Neurophysiol, 63, pp 82-104, 1990).

7. F Crick and C Koch, *The Problem of Consciousness* (Scientific American, September, 1992); see also F Crick and C Koch, *The Hidden Mind* (ibid, August, 2002).

8. For a discussion, see http://waynesword.palomar.edu/termfl1.htm

9. http://www1.umn.edu/news/news-releases/2001/UR_RELEASE_MIG_1251.html

10. http://www.beep.ac.uk/content/116.0.html

11. I Klimanskaya, Y Chung, S Becker, S-J Lu and R. Lanza, *Human embryonic stem cells derived from single blastomeres* (Nature, vol 444, p 481, 2006).

12. M Nakagawa et al, *Generation of induced pluripotent stem cells without Myc from mouse and human fibroblasts* (Nature Biotechnology, vol 26, pp 101-106, 2008).

Chapter 10: Predictions and Feedback

1. R P Feynman, *Surely You're Joking Mr. Feynman!* (W W Norton and Co, New York, 1985).

2. I R Morus, *Michael Faraday and the Electrical Century* (Icon Books Ltd, 2004).

3. For a popular account, see J Gleick, *Chaos* (Penguin, New York, 1987).

4. B B Mandelbrot, *Fractal Geometry: what is it, and what does it do?* (Proceedings of the Royal Society of London, vol 243, pp 3-16, 1989).

5. http://www.mcs.surrey.ac.uk/Personal/R.Knott/Fibonacci/fibnat.html#bees

6. M Ronan, *Symmetry and the Monster – One of the Greatest Quests of Mathematics* (Oxford University Press, 2006).

7. http://library.thinkquest.org/26742/hitler.html

8. J E Lovelock, *Gaia – A New Look at Life on Earth* (Oxford University Press, 1979 and 1987).

9. *The Revenge of Gaia: Why the Earth Is Fighting Back — and How We Can Still Save Humanity* (Perseus, 2006).

Chapter 11: Behaviour

1. R Dawkins, *The Selfish Gene* (Oxford University, Oxford, UK, 1976).
2. E O Wilson, *Sociobiology – the Abridged Edition* (The Belknap Press of Harvard University Press, Cambridge, Mass, 1980).
3. P G Bateson in *Evolution from Molecules to Men* (D E Bendall, Ed, Cambridge University Press, 1983).
4. J Alcock, *Animal Behavior: An Evolutionary Approach* 9th Edition (Sinauer Publishers, 2009).
5. J R Krebs and N B Davies, *An Introduction to Behavioural Ecology* (Blackwell, Oxford, 1987).
6. N Tinbergen, *The Study of Instinct* (Oxford University Press, Oxford, 1989).
7. R Dawkins, *The Extended Phenotype: the Long Reach of the Gene* (Oxford University Press, Oxford, 1999, 2nd Edition).
8. N J Emery and N S Clayton, *The Mentality of Crows: Convergent Evolution of Intelligence in Corvids and Apes* (Science, 306, pp 903-1907, 2004).
9. D H Williams, E Stephens and M Zhou, *Ligand Binding Energy and Catalytic Efficiency from Improved Packing within Receptors and Enzymes* (J Mol Biol, vol 329, pp 389-399, 2003).
10. D H Williams, M Zhou and E Stephens, *Ligand Binding Energy and Enzyme Efficiency from Reductions in Protein Dynamics* (J Mol Biol, vol 355, pp 760-767, 2006).
11. S Bowles and H Gintis, 2002, *dahlem.pdf* from http://www.umass.edu
12. R L Day, T MacDonald, C Brown, K N Laland and S M Reader, *Interactions between shoal size and conformity in guppy social foraging* (Animal Behaviour, 63, p 193, 2002).
13. J Diamond, *Guns, Germs and Steel* (Jonathan Cape, London, 1997).
14. J Maynard Smith, *Evolution and the Theory of Games* (Amer Sci, vol 64, p 41, 1976).
15. http://www.psychol.cam.ac.uk/pages/staffweb/clayton/Scrub_jays.html
16. http://en.wikipedia.org/wiki/Bhopal_disaster
17. http://en.wikipedia.org/wiki/Bloody_Sunday_(1972)
18. D Morris, *The Naked Ape; a Zoologist's Study of the Human Animal* (McGraw-Hill, New York, 1st US edition, 1967).
19. D Morris, *The Human Zoo* (McGraw-Hill, New York, 1969).
20. BBC Natural History Unit, *Life* (a televised Nature Documentary Series, 2009).

Chapter 12: Why Freedom Dies

1. G Strawson, *The Impossibility of Moral Responsibility* (Philosophical Studies, vol 75, pp 5-24, 1994).
2. R Penrose, *The Emperor's New Mind* (Vintage, London, 1990).
3. http://www.dana.org/news/brainhealth/detail.aspx?id=10062
4. H Stapp, *Quantum theory and the role of the mind in nature* (Foundations of Physics, vol 31, pp 1465-1499, 2001).
5. S Klein, *Libet's research on the conscious intention to act. A commentary* (Consciousness and Cognition, vol 11, pp 273-279, 2002).
6. J Humphrys, *What I found out about God* (The Daily Telegraph, London, 23 December 2006).
7. B F Skinner, *Science and Human Behavior* (Macmillan Publishing Inc, New York, 1953).
8. P Gorniak, Meaning 'I' (*www.media.mit.edu/cogmac/.../gorniak_generals_paper.pdf*)
9. See, for example, www.cambridgewellbeing.org
10. S Volkov, *Testimony – the Memoirs of Dmitri Shostakovich* (Hamish Hamilton Ltd, London, 1979).
11. E O Wilson, *Sociobiology – the Abridged Edition* (The Belknap Press of Harvard University Press, Cambridge, Mass, 1980).
12. C M Breder Jr and C W Coates, (Copeia, vol 3, pp 147-155, 1932).
13. S D Levitt and S J Dubner, *Freakonomics: A Rogue Economist Explores the Hidden Side of Everything* (Allen Lane, Great Britain, 2005).
14. R V Jones, *Most Secret War* (Hamish Hamilton Ltd, 1978).
15. I Finkelstein and N A Silberman, *The Bible Unearthed* (Touchstone, New York, NY, 2002).
16. R Dawkins, *The Selfish Gene* (Oxford University Press, 1989 edition, chapter 11).
17. S Blackmore, *The Meme Machine* (Oxford University Press, 1999).
18. S E Asch, *Opinions and Social Pressure* (Scientific American, vol 193, pp 31-35, 1955).

Chapter 13: Humanity Illustrated

1. Moses I Finley, *Ancient History – Evidence and Models* (Viking Penguin Inc, 1986).
2. Alison Weir, *Eleanor of Aquitaine* (Jonathan Cape Ltd, 1999).
3. Jonathan Phillips, *The Fourth Crusade and the Sack of Constantinople* (Jonathan Cape Ltd, 2004).
4. A N Wilson, *The Victorians* (Arrow Books, 2003).

5. Robert Service, *Stalin – a Biography* (Macmillan, 2004).

6. Jung Chang and Jon Halliday, *Mao – the Unknown Story* (Jonathan Cape, 2005).

7. Lindsey Hughes, *Peter the Great – a Biography* (Yale University Press, 2002).

8. Henri Troyat, *Catherine the Great* (Phoenix Press, 1979).

9. Nancy Mitford, *Frederick the Great* (Penguin Books Reprint Edition, 1995).

10. Vincent Cronin, *Napoleon* (Harper Collins, London, 1971).

11. Christopher Hibbert, *Wellington* (Harper Collins, 1997).

12. Terry Coleman, *Nelson, The Man and the Legend* (Bloomsbury, London, 2001).

13. Alun Chalfont, *Montgomery of Alamein* (Atheneum, New York, NY, 1976).

14. David Lodge, *Thinks* (Secker and Warburg, UK, 2001).

15. Joann Hames and Yvonne Ekern, *Introduction to Law* 4th Edition (Prentice Hall, 2009).

16. Statement by the U S President, White House News Release, April 2004 (http://www.whitehouse.gov/news/releases/2004/04/20040419-4.html). In relation to Guantanamo Bay, see also http://en.wikipedia.org/wiki/Guantanamo_Bay

17. Extraordinary rendition see http://www.washingtonpost.com/wp-dyn/content/article/2005/12/05/AR2005120500240.html

18. http://www.bailii.org/ew/cases/EWCA/Civ/2009/441.html

19. E O Wilson, *Consilience – the Unity of Knowledge* (Little, Brown and Co, UK, 1998).

20. http://www.americanthinker.com/blog/2010/01/defending_the_rights_of_the_un.html

21. I Chang, *The Rape of Nanking* (Basic Books, USA, 1997).

22. J Fenby, *Generalissimo Chang Kai-Shek, and the China He Lost* (The Free Press, UK, 2003).

23. P E Lovejoy, *Transformations in Slavery: A History of Slavery in Africa* (Cambridge University Press, UK, 1983).

24. N Ferguson, *Empire – How Britain Made The Modern World* (Penguin Books UK, 2004).

25. C Andrew, *The Defence of the Realm, the Authorized History of MI5* (Penguin Books, 2009, p 795).

26. R M Axelrod, *The Evolution of Cooperation* (Basic Books, New York, 1984).

27. M A Novack, R M May and K Sigmund, *The Arithmetics of Mutual Help* (Scientific American, June, 1995, pp 76-81).

28. A I Solzhenitsyn, *Cancer Ward* (Penguin Books, UK, 1971).

29 .A I Solzhenitsyn, *One Day in the Life of Ivan Denisovich* (Penguin Classics, UK, 2000).

30. J Boyne, *The Boy in the Striped Pyjamas* (Random House Children's Books, London, 2006).

Chapter 14: Humanity Tested

1. P G Zimbardo and F L Ruch, *Psychology and Life* (Scott, Foreman and Co, Glenview, Illinois, 1975).

2. C Haney, C Banks and P Zimbardo, *Interpersonal Dynamics in a Simulated Prison* (Int J Criminology and Penology, no 1, p 73, 1973).

3. M Gladwell, *The Tipping Point* (Abacus Books, 2000).

4. For a discussion of the biological bases of morality, see R A Hind, *Why Good is Good – the Sources of Morality* (Routledge, London, 2002).

5. J Darley and D Bateson, *From Jerusalem to Jericho: A study of situational and dispositional variables in helping behavior* (J Personality and Soc Psychol, vol 27, pp 100-119, 1973 cited in M Gladwell, *The Tipping Point* Abacus Books, 2000).

6. http://www.ralentz.com/old/space/feynman-report.html

7. R P Feynman, *Surely you're joking, Mr Feynman!* (W W Norton and Co, New York, NY, 1985).

8. B Libet, *Mind Time* (Harvard University Press, Cambridge, Mass, USA, 2004).

9. D Ariely, *Predictably Irrational* (Harper Collins, revised edition, 2009).

10. J B Moseley et al, (New England Journal of Medicine, vol 347, pp 81-88, 2002).

11. J-K Zubieta et al, (J Neuroscience, 24 August 2005).

Chapter 15: The Human Condition

1. For insights into some of the plays, see J Shapiro, *1599 – A Year in the Life of William Shakespeare* (Faber and Faber Ltd, London, 2005).

2. B W Tuchman, *The March of Folly: From Troy to Vietnam* (Ballantine Books, New York, 1984).

3. Opinion of Alan Greenspan, see http://news.bbc.co.uk/1/hi/8244600.stm

4. *The Independent*, edition of 19 October 2007.

5. http://www.thefreedictionary.com/IQ

6. http://news.bbc.co.uk/1/hi/sci/tech/7124156.stm

7. http://www.london.gov.uk/view_press_release.jsp?releaseid=14115

8. House of Commons Science and Technology Committee, *Scientific*

Evidence, Risk and Evidence Based Policy Making The Stationery Office Ltd, UK, 8 November 2006.

9. http://www.telegraph.co.uk/news/newstopics/politics/health/2969039/Chemical-in-food-tins-doubles-the-risk-of-heart-disease-and-diabetes.html

10. *Chemistry World* (Royal Society of Chemistry, UK, p 6, September 2008; and p 13, October 2008).

11. http://www.worldviewofglobalwarming.org/pages/glaciers.html

12. For the majority scientific view on the causes of global warming in 2001, see http://www.ncdc.noaa.gov/oa/climate/globalwarming.html (mainly a summary of the IPCC 2001 report *Climate Change 2001: The Scientific Basis*).

13. D Evans, *Global Warming or Cooling? A New Trend in Climate Alarmism* http://www.globalresearch.ca/index.php?context=va&aid=14504

14. D C MacKay, *Sustainable Energy – without the hot air* (UIT, Cambridge, England, 2009).

15. D Lawson, *The Independent*, 8 June 2007.

16. B Goldacre, *ad Science* (Fourth Estate, 2008).

Chapter 16: Humanity Challenged

1. S Rose, R C Lewontin and L J Camin, *Not in our Genes* (Pelican Books, 1984).

2. E J Larson and L Witham, *Leading scientists still reject God* (Nature, 394, p 313, 1998).

3. Lewis Wolpert, *Six Impossible Things before Breakfast: The Evolutionary Origins of Belief* (Faber, UK, 2006).

4. Jonathan Miller, *A Brief History of Disbelief* www.bbc.co.uk/bbcfour/documentaries/features/atheism.shtml

5. http://www.adherents.com/Religions_By_Adherents.html

6. R A Fisher, *The Genetical Theory of Natural Selection* (Clarendon Press, Oxford, 1930).

7. R Dawkins, *The Blind Watchmaker* (Longman, Harlow, UK, 1986).

8. A Kornberg, *The Two Cultures: Chemistry and Biology* (Biochemistry, 26, 6888, 1987).

9. *Time* magazine, December 1993; see also http://demons.monstrous.com/existence_of_the_devil.htm

10. T Wilkinson, *The Vatican's Exorcist* (Warner Books Inc, 2007).

11. P W Atkins, *The Creation* (W H Freeman, Oxford, 1981).

12. http://www.parliament.the-stationery-office.co.uk/pa/cm199697/cmhansrd/vo961211/debtext/61211-49.htm

13. T Fahey, A A Montgomery, J Barnes and J Protheroe, *Quality of care for elderly residents in nursing homes and elderly people living at home: controlled observational study* (BMJ, vol 326, p 580, 2003).
14. P Johnston, *The old and the sick shouldn't be given a quick exit button, The Daily Telegraph,* 10 October 2005.
15. Website of the Christian Medical Fellowship: http://www.cmf.org.uk/index/joffe_bill.htm
16. http://www.adherents.com/misc/religion_suicide.html
17. http://www.un.org/News/Press/docs/2008/sgsm11451.doc.htm
18. S Hawking and L Mlodinow, The Grand Design (Bantam Press, 2010).

Chapter 17: Big Questions

1. R Keynes, *Fossils, Finches, and Fuegians* (Harper Collins, London, 2002).
2. P Medawar, *The Limits of Science* (Oxford University Press, 1986).
3. P Davies, *The Mind of God* (Penguin Books, 1993).
4. J Monod, *Chance and Necessity* (Knopf, New York, 1971).
5. A Weisman, *The Selection Theory, in 'Darwin and Modern Science'* (A C Seward ed, Cambridge University Press, 1909).
6. E Mayr, *Towards a New Philosophy of Biology* (Belknap Press of Harvard Univ Press, Cambridge, Mass, 1988).
7. L M Foster, (Scientific American, October 1995, p 17).
8. B Bryson, *A Short History of Nearly Everything* (Doubleday, UK, 2003).
9. http://www.timesonline.co.uk/tol/news/politics/article2980432.ece
10. http://www.vatican.va/holy_father/benedict_xvi/encyclicals/documents/hf_ben-xvi_enc_20071130_spe-salvi_en.html
11. http://www.avert.org/subaadults.htm
12. http://www.harrisinteractive.com/harris_poll/
13. J D Miller, E C Scott and S Okamoto, *Public Acceptance of Evolution* (Science, 11 August 2006, pp 765-766); see also http://newsroom.msu.edu/site/indexer/2827/content.htm
14. R Dawkins, *The God Delusion* (Bantam Press, London, 2006).
15. R Dawkins, *Unweaving the Rainbow: Science, Delusion and the Appetite for Wonder* (Penguin, 1998).
16. C Hitchens, *God is not Great* (Atlantic Books, London, 2008).
17. D C Dennett, *Breaking the Spell: Religion as a Natural Phenomenon* (Viking, 2006).
18. E O Wilson, *Consilience – the Unity of Knowledge* (Little, Brown and Co, UK, 1998).
19. http://www.un.org/News/Press/docs/2008/sgsm11451.doc.htm

20. L Trut, I Oskina, A Kharlamova, Animal evolution during domestication: the domesticated fox as a model (Bioessays, vol 31, pp 349–360, 2009).
21. L N Trut, *Early Canid Domestication: the Farm-Fox Experiment* (American Scientist, vol 87, pp 160–169, 1999).
22. http://www.bbc.co.uk/iplayer/episode/b00j1bhw/b00j1b9p/Natures_Great_Events_The_Great_Tide/
23. http://www.unicef.org/sowc08/index.php

Index

photosynthesis
  cells and energy 65
  Gaia hypothesis 160–1
  global warming 236
  quantum mechanics 55–7
Picasso, Pablo 92
placebo effect 226–7, 249
plagues 159
Planck, Max 51
plate tectonics 41
Plato 249
playing God 144–5
Poincaré, Jules Henri 19
Pol Pot 199
politics
  humanity 210–14
  science and 231–4
  truths and fictions 2–3
Popper, Karl 1, 257
populations 151–6
positive feedback 159–60, 170
predictions 146–61
  boring predictions 146
  branching systems 157
  cause and effect 4
  feedback control 150–1, 154–6, 159–61
  Gaia hypothesis 160–1
  mathematics 154–8
  media and science 236
  populations 151–6
  societies and the ratchet effect 159–60
  testing for fitness 148
  what happens next? 146–8
predictive economics 206–7
Priestley, Joseph 210
probability
  false positives 227
  migration 106
  quantum mechanics 50
  truths and fictions 1–3
  of the Universe 45–6
probable truths 2–3
prokaryotic cells 65
proteins
  cooperation 171, 173
  disease 131–2
  evolution of organisms 98–9
  genetic code 70–1
  natural selection 82, 84, 91, 93

structure and function 71–4
pseudo-science 232, 235
psychology 226–7
pulsars 28
punishment 216–19, 221–3

quantitative easing 208
quantum mechanics 47–60
  assembling organisms 56–7
  Boltzmann, Ludwig 50–1
  change and equilibrium 57–9
  direction of change 53–5
  entropy 50–1, 54–5
  freedom 182
  localised energy sources 55–6
  lumpiness of energy 48–9, 51
  natural frequencies of particles and
    aggregates 52–3
  queerness of small things 47–8
  requirements for life 59–60
  uncertainty principle 47–8
quarks 19–20

race and racism 229–30
racial cleansing 175
radio waves 146, 149
radioactive decay 39–40
Ranke, Leopold von 197
Rape of Nanking 217
Rasputin 249
ratchet effect 159–60
rational markets 155
recycling 209–10
red shift 10
relativity, general theory of 8
religion
  creation and the Big Bang 255
  creationism and intelligent design 242–4
  crusades 197–8
  Earth's solar system 38
  freedom 180, 182, 187–8, 193–4
  genetic engineering 141
  mindless forces and humanity 261–2
  science and 244–8
rendition 205
reparations 217–18
reproducibility 4
reproduction 67
  mechanical tinkering 140

www.ingramcontent.com/pod-product-compliance
Lightning Source LLC
Chambersburg PA
CBHW031501270326
41930CB00006B/186